当代水墨艺术的"正发生"
AI and Painting: The "Happening" of Contemporary Ink Art

海洋出版社
2024年·北京

作者简介

凡悲鲁，动画家、漫画家、当代艺术家，师从艺术家方明（持永只仁）、阿达（徐景达）、方成、韩羽等。享受国务院特殊津贴，中国美术家协会理事，出版画集有《凡悲鲁彩墨画集》《江湖》《人心》《凡悲鲁画册》《墨迹》《墨像》《墨途》等，其漫画作品被美国、韩国等多个国家的高校及文化机构和个人收藏。

其笔名"凡悲鲁"三字分别代表三个人物，"凡"代表凡高，"悲"代表徐悲鸿，"鲁"代表鲁迅，这三个伟大的人物是在凡悲鲁少年时期对他产生巨大影响的艺术家。凡悲鲁希望能够汲取他们身上的艺术精神，创作出具有思想性、艺术性、观赏性的作品。

About the Author

Fanbeilu, a well-known comic artist and contemporary artist once studied under artists such as Fang Ming (Tadahito Mochinaga), Ada (Xu Jingda), Fang Cheng, and Han Yu. Currently he is a member of China Artists Association and the director of the Animation Arts Committee. He also enjoys the special allowance of the State Council. His representative comic works includes: *The Color Ink Painting Collection of Fanbeilu*, *Jiang Hu*, *Ren Xin*, *The Illustrations of Fanbeilu*, *Traces in Ink*, *Faces in Ink*, *Journey in Ink* etc, many of these works had been collected and exhibited by individuals, academies and cultural institutions from foreign countries. His preliminary design of animated short films *Qiu Shi*, *Li Qiu*, and *Xin Sanchakou* were displayed at the 2nd Multi-Media Art Exhibition in 2020.

"Fan" refers to Vincent Willem van Gogh, "Bei" refers to Xu Beihong, and "Lu" refers to Lu Xun. Each of them had brought enormous impact to his understanding of art during his teenage years. Fanbeilu shows strong sincerity towards their artistic and spiritual expression in his own art experiments that strives to create works with high standard of ideological, artistic and ornamental quality.

序

比起技术应用说明书，我更希望启发关于"美"的思考

在刚刚过去的2023年，我们见证了人工智能生成绘画技术突飞猛进的发展。围绕"AI（人工智能）""绘画艺术""媒介融合"等关键词刊出的图书与期刊比比皆是，在网页搜索栏中检索"AI绘画"，可以轻松找到种类繁多的"极简咒语公式（prompt）"以生成特定风格样式的图片。在这一"AI绘画"热潮的反面，则多为艺术行业从业者对于行业未来发展前景的隐忧，"人工智能是否会在艺术创作方面取代人类"成为被讨论的热点。但纵览海量"前沿"资讯，却鲜少在琳琅满目的"使用教学""应用教程""速成方法"中找到有关于绘画之"美"本身的讨论，鲜少有观点能够呈现出对"生成式"绘画审美范式背后的地域文化底色、视觉思维逻辑等的辩证思考。因此比起推出一本技术应用说明书，我更希望启发大众对"美"产生自主性的思考。

本书所收录的内容既有我日常即兴绘制的水墨手稿，又有基于我的水墨手稿数据库，借助人工智能技术生成的图稿。我提供给人工智能的作品在技法与媒材应用上都具有鲜明的中国民族性，如"散点透视""墨分五彩"等，但也在构图、造型、空间、视点中融入了一定的西方美术思维。在十多位同学的辅助下，我们将这些绘画作品扫描，建立了初具规模的图像数据库。"基于中国审美的图像数据库，人工智能究竟能识别出什么？又能生成出什么？"是本书讨论的出发点，进而我希望同读者共同深入思考"美的思维、美的建构、美的差异是否可以被人工智能所习得？""人工智能时代的艺术创作中，人与信息、人与算法、人与自我之间的关系究竟为何？"等关键问题，最终我希望通过这次颇具实验性的尝试，激励当代艺术创作者活用人工智能技术，人机协同，以原创力为核心驱动力，更好地讲述中国故事、传递中国声音、彰显中国审美。

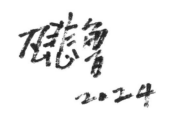

基于右图的 AI 衍生数字绘画　　　　实验水墨
媒材模拟：复印纸，水墨　　　　　　绘画材料：复印纸，水墨

AI-derived digital painting based on the painting (Right)　　Experimental Ink Wash Painting
Simulated Material：Duplicating paper, Ink　　Artwork：Duplicating paper, Ink
928px×1232px　　　　　　　　　　33cm×46cm
2023　　　　　　　　　　　　　　　2023

I wish to inspire fresh thoughts on "beauty" rather than providing a technical manual.

In 2023, we witnessed the rapid development of artificial intelligence-generated painting technology. Books and journals featuring keywords such as "AI (Artificial Intelligence)" "painting art" and "media integration" abound. A simple search for "AI painting" easily yields a variety of "minimalist incantation formulas (prompts)" that can be used to generate images in specific styles. On the flip side of this "AI painting" trend, there are concerns among art industry professionals about the future prospects of the industry, with the hot topic being whether artificial intelligence will replace humans in artistic creation. However, despite the abundance of "cutting-edge" information, discussions about the "beauty" of painting itself are scarce. There is a lack of exploration of the cultural background and visual logic behind the "generative" aesthetic paradigm in the midst of the overwhelming array of "usage tutorials" "application guides" and "quick learning methods". Therefore, rather than releasing a technical manual, I hope to inspire the public to autonomously contemplate the inner "beauty" of paintings, both the hand drawn ones and the AI generated ones.

实验水墨
绘画材料：宣纸板，水墨

Experimental Ink Wash Painting
Artwork：Rice paper board, Ink
33cm×46cm
2024

基于左图的 AI 衍生数字绘画
媒材模拟：宣纸板，水墨

AI-derived digital painting based on the painting (Left)
Simulated Material：Rice paper board, Ink
928px×1232px
2024

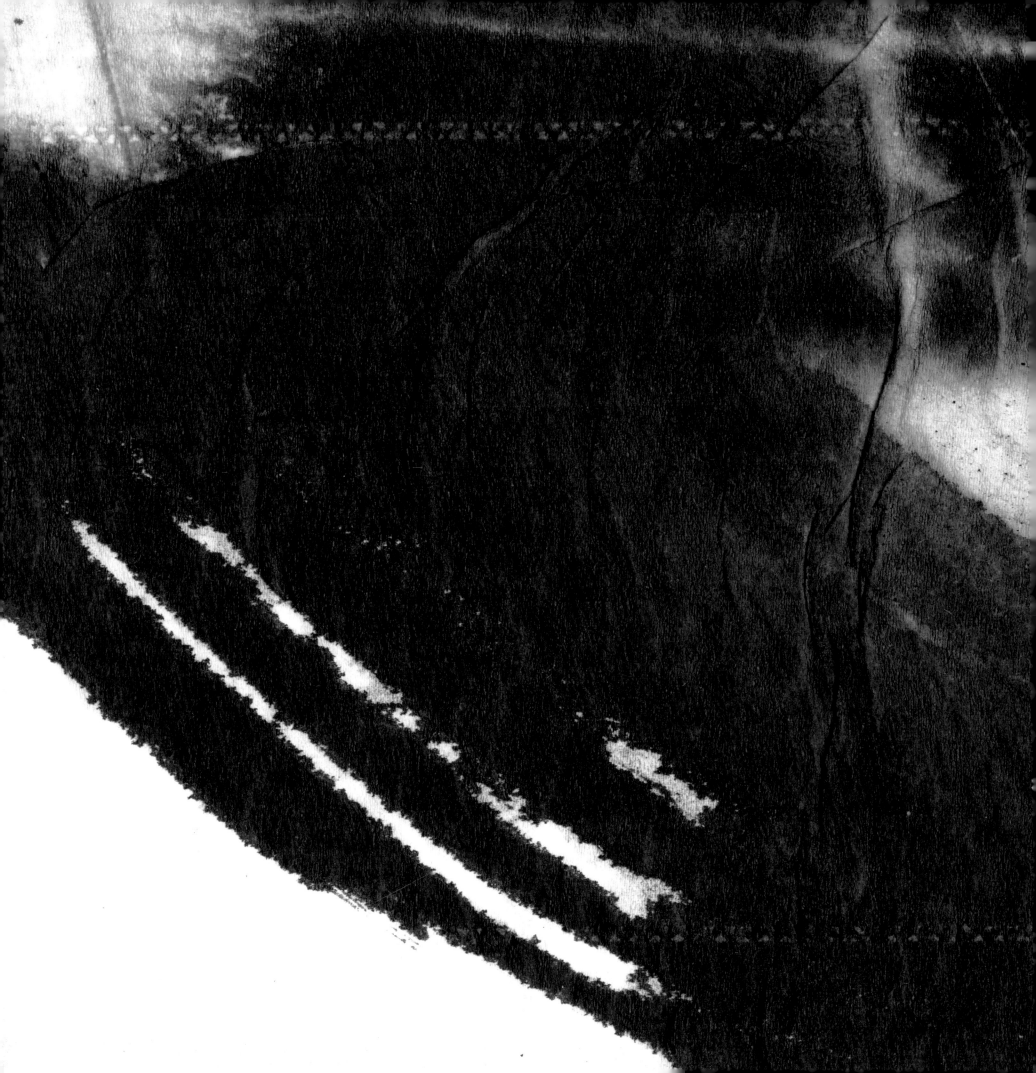

目录
CONTENTS

002 第一章 / Chapter 1
落笔：体悟传统笔墨绘画的质料性
Brushstroke: Understanding the Materiality of Traditional Ink Painting

086 第二章 / Chapter 2
临场：见证水墨艺术未知的多样性
On Scene: Witnessing the Unknown of Ink Art

172 第三章 / Chapter 3
碰撞：认知人工智能艺术的生成性
Collision: Recognizing the Generative Quality of AI Art

258 第四章 / Chapter 4
发现：碰撞数字绘画艺术的可能性
Discovery: Exploring the Possibilities of Digital Painting Art and AI

332 构建人工智能时代的东方审美
Construct the Oriental Aesthetic in the Era of Artificial Intelligence

348 后 记 / Afterword

第一章
落笔：体悟传统笔墨绘画的质料性

Chapter 1

Brushstroke: Understanding the Materiality of Traditional Ink Painting

第一章 落笔：体悟传统笔墨绘画的质料性

采访01："无纸化"在今天似乎已然成为数字艺术生产的大势所趋。但手绘似乎是您一直以来内化于生活日常的习惯，请问手绘于您而言的"不可替代性"为何？

凡悲鲁（下文简称凡）：我是在1992年被派去香港学习计算机三维动画技术，一共派出去四个人，我是其中唯一一个动画专业出身的，也正是得益于动画专业学习所带来的形象思维惯性，我在初步接触三维动画技术时相对要得心应手。但当时给我的感受是计算机动画与我所理解的动画相较甚远，它更强调对"计算机语言"的熟练掌握，并不强调动画造型与视听的艺术性。并且当时所谓的三维，多是借助摄像机的景深、多层台的拍摄，以木偶的方式实现的，更贴合特技摄影，而非真正意义上的三维动画。当然伴随着技术日新月异的发展，现在我们可以在全然"无纸化"的条件下完成二维、三维乃至VR、AR动画制作。然而对我来说，所谓的"绘画"并不只是被他者观看的那一幅画，虽然画本身固然重要，但绘画的初心与绘画的过程是我更为珍视的。

绘画的初心很简单，就是单纯地喜欢。如何定义这种"喜欢"，大概就是相对于其他日常事务而言，比如以半天为单位的授课、大大小小的会议、长篇大论的文字撰写等。我在作画的时候，无论花费多长的时间，不会觉得辛苦，或者说乐此不疲更为贴切。绘画的过程亦是充满了未知的乐趣，比如我会刻意不清洗毛笔和砚台，砚台里的墨干掉后，再加水稀释，再干掉，再稀释……如此几个回合后，你会发现笔墨所呈现出的效果与初始状态大相径庭，色泽浓度的差异是显而易见的，此外更有浓墨干涸后遗留的残渣，当你用毛笔将其混合着稀释后的墨汁涂到不同的材料上时，经常会留下一些近似于粉尘杂质一样的东西，这便是"不期而遇"的"不可替代"。借由笔触皴擦点染的力道与章法，我参与了水、墨、纸、笔质性的碰撞，这一过程是充满乐趣与惊喜的。

Chapter 1　Brushstroke: Understanding the Materiality of Traditional Ink Painting

实验水墨
绘画材料：速写纸，水墨

Experimental Ink Wash Painting
Artwork:Sketching paper, Ink
21cm×28cm
2023

第一章　落笔：体悟传统笔墨绘画的质料性

Q1: "Paperless production" seems to have become a trend in the making of digital artworks today. However, hand-drawing appears to be a habit that has internalized into your daily life. What do you think is the irreplaceable quality of hand-drawing?

实验水墨
绘画材料：速写纸，水墨

Experimental Ink Wash Painting
Artwork: Sketching paper, Ink
10cm × 16cm
2023

Fan: I was sent to Hong Kong to study 3D computer animation technology in 1992. Among the four people sent, I was the only one majored in animation. It was precisely due to the image thinking habits brought about by studying animation that I felt relatively not that challenging but rather intrigued when I initially encountered 3D animation technology. However, at that time, my impression was that computer animation was quite different from my understanding of animation. It emphasized proficiency in the "computer language" rather than emphasizing the artistic aspects of animation modeling and audio-visual effects. The so-called three-dimensionality at that time often relied on the depth of field and multi-layered shooting with a camera, implemented in a puppet-like manner, which was more in line with special effects photography rather than true 3D animation. Of course, with the rapid development of technology, we can now complete 2D, 3D, and even VR and AR animation production under completely "paperless" conditions. However, for me, the so-called "painting" refers to not just the visual image that others observe. While the visual image is important, the original intention of painting and the process of painting are what I cherish more.

My initial intention of painting is very simple—it's purely out of love. How to define this "love" is probably in comparison with other daily affairs. For instance, activities like teaching for half

Chapter 1 Brushstroke: Understanding the Materiality of Traditional Ink Painting

a day, countless meetings, and lengthy writing etc. When I'm painting, regardless of how much time it takes, I don't feel it's laborious, in fact, I find it enjoyable and tireless. The process of painting is filled with unknown pleasures. For example, I deliberately skipped cleaning the brush and ink stone. After the ink in the ink stone dries up, I add water to dilute it, let it dry again, dilute it once more, and so on. After a few rounds, you'll notice a stark difference in the effect presented by the ink compared to its initial state. The visible difference in color density is apparent. Additionally, there are residues left behind after the thick ink dries, and when you use the brush to mix it with the diluted ink and apply it to different materials, you often leave behind something resembling dust or impurities. This is the "unexpected encounter" and the "non-duplicated." Through the strokes, rubbings, and dyes applied with the force and methods of brushwork, I engage in the materialized collision of water, ink, paper, and the brush. This process is filled with joy and surprises.

第一章 落笔：体悟传统笔墨绘画的质料性

采访02：您在绘画媒材的选择上最看重的是什么？

实验水墨
绘画材料：复印纸，水墨

Experimental Ink Wash Painting
Artwork:Duplicating paper, Ink
21cm×28cm
2023

凡：我在绘画媒材上的选择具有高度的"不确定性"，这是从学生时代起便有的习惯，即不纠结于媒材的品质高低与媒材间"约定俗成"的适配度，而更注重表现特定场景、特定人物所传达出的氛围、情绪等。记得1985年，周萍老师带我们到苏州去写生，晚上会讲评作业，当时周萍老师请来了他的老师——吴冠中先生来给我们讲画、评画。以前我们都更多用的是铅笔、钢笔、炭笔来画速写，而我是直接用毛笔画的，画在当地产的马粪纸或毛边纸上，是一辆写生地随处可见的双轮小推车。我记得那幅画当时还被吴冠中老师表扬了，虽然毛笔很难在毛边纸上勾勒细节，但我在画风和构图上呈现出那个场景应有的氛围，这在吴冠中老师看来是很难得的。包括在这本书收录的画中，我所用的往往是我在"想要作画"的那个当下信手拈来之物，有些甚至很难称得上是画材。

古语称"纸寿千年"，在纸张上绘画总有一种电脑绘画无法取代的厚重感。我所使用的有画框衬纸、信封信纸、牛皮纸公文夹甚至是餐巾纸，晕开干墨的水有时是手边的一杯茶……但也恰恰是因为使用了在灵感产生的当下眼睛所捕捉到的物，哪怕是隔了很长一段时间，对于绘制那幅作品的记忆仍然可以是鲜活灵动的，某种意义上有点接近"图像日志"的概念，画作本身凝结了我在特定环境下的所感、所思、所想，这种特质在我看来是弥足珍贵的。

Chapter 1 Brushstroke: Understanding the Materiality of Traditional Ink Painting

实验水墨
绘画材料：复印纸，水墨

Experimental Ink Wash Painting
Artwork:Duplicating paper, Ink
21cm×28cm
2023

Q2: What do you value the most in your choice of painting materials?

Fan: My choice of painting materials is highly uncertain, a habit I developed since my early years as an art student. I don't get bogged down by the quality of the materials or their conventional compatibility. Instead, I focus more on expressing the atmosphere and emotions conveyed by specific scenes and characters. I remember in 1985, our teacher Zhou Ping took us to Suzhou for outdoor sketching. During the critique sessions which always took place at night, Zhou Ping invited his teacher, Wu Guanzhong, to evaluate our paintings. At that time, most of us used pencils, ballpoint pens, or charcoal to do sketches, while I used a brush directly, drawing on locally produced horse dung paper or rough-edged paper, I remember vividly drawing a small two-wheel cart easily found at the sketching site. I recall that Wu Guanzhong praised that piece. Although it was challenging to capture details with a brush on rough-edged paper, I presented the atmosphere and composition of the scene appropriately, which he found to be more remarkable.

Including the paintings featured in this book, I often use whatever is readily available at the moment I want to paint, some of which can hardly be considered traditional art supplies. The ancient saying goes, "Paper lasts a thousand years," and there's a weightiness to painting on paper that digital painting cannot replace. I use framed paper, postal code letter paper, parchment document folders, and even napkins. The water I use to dilute ink sometimes comes from a cup of tea at hand. However, sometimes precisely because I use something captured by my eyes in the moment of inspiration, even after a long time, the memory of creating that work can still be vivid and dynamic. In a sense, it's akin to the mechanism of a "visual diary". The artwork itself encapsulates my feelings, thoughts, and ideas in a specific environment, a quality I find immensely precious.

实验水墨
绘画材料：宣纸，水墨

Experimental Ink Wash Painting
Artwork:Rice paper, Ink
33cm×46cm
2023

采访03：可否这样理解，绘画媒材的"不确定性"也是您即兴创作的一部分？

凡： 在某种意义上是的。不仅限于绘画的媒材，包括我有时会即兴以国画的形式，但使用西画的材料，甚至在构图上组合东西方的视觉语言等等。我在个人的创作中，无论是真人实拍，还是动画抑或是写生、速写等等，我都特别强调"即兴"的意义，这种偶发的艺术灵感对我来说特别重要。哪怕是在户外极限运动挑战的过程中，比如跑半马、越野赛时，我也要将速写本随身携带。我想通过我自己的实际参与，来探索人在物理意义上的极端状态下的艺术感知与日常状态有何不同。他眼睛所捕捉到的画面，他落笔所呈现的画面将如何与其身心状态产生呼应：比如我先无间歇地跑上六个小时，然后再拿出笔来，甚至可以不在速写本上，可以在树叶、树干、路边被丢弃的饮料瓶、易拉罐等物上画下目之所及处激发我灵感的景致……同样的景致，如果被拍成照片，给到常年在画室里画静物的艺术家们，或许完全无法成为"值得被描摹"的客体，因为他们是以理性的目光将其加以审视，而处于身心极度疲劳状态下的我则对其倾注了更多的、更为鲜活的感性，这些感性的因素都是不期而遇的、不确定的，却也是能够激活一幅画作艺术灵韵的存在。我不知道是否有其他艺术家在做这类的实验，我自己已经完成了11本速写，将来这些作品将被收录至《边跑边画》一书中，届时我也希望画作中的某种"不确定性"能够与读者产生心灵层面的共振。

Chapter 1　Brushstroke: Understanding the Materiality of Traditional Ink Painting

实验水墨
绘画材料：牛皮纸，水墨

Experimental Ink Wash Painting
Artwork: Kraft paper, Ink
21cm × 28cm
2023

实验水墨
绘画材料：牛皮纸，水墨

Experimental Ink Wash Painting
Artwork: Kraft paper, Ink
21cm × 28cm
2023

Q3: So the "uncertainty" of painting materials can be understood as part of your spontaneous creation?

Fan: In a sense, yes. Not limited to the painting materials, I sometimes spontaneously adopt the form of traditional Chinese painting but use Western painting materials or combine visual patterns from both East and West in composition. In my personal creations, whether it's real-life photography, animation, sketches, or other art forms, I emphasize the significance of "spontaneity". These incidental artistic inspirations are particularly important to me. Even in the process of extreme outdoor sports challenges, such as running a half marathon or participating in cross-country races, I carry a sketchbook with me. I aim to explore how artistic perception in extreme physical states differs from daily states through my actual participation. How the scenes captured by my eyes, the images presented by my brush, resonate with my physical and mental state: for example, running continuously for six hours before taking out a pen. I may not use a sketchbook; I might draw on leaves, tree trunks, discarded beverage bottles, or cans by the roadside—whatever inspires me. The same scene, if captured in a photograph and given to artists who typically paint still life in their studios, might not be deemed "worthy of depiction" because they analyze it with a rational gaze. In my extreme state of physical and mental fatigue, I infuse more vivid and sensory elements into it. These sensory factors are unexpected and uncertain, yet they can activate the artistic charm of a painting. I don't know if other artists are conducting similar experiments, but I have completed 11 sketchbooks. In the future, these works will be included in the book *Sketching While Running*. At that time, I hope that a certain "uncertainty" in the artworks can resonate with readers on a spiritual level.

Chapter 1 Brushstroke: Understanding the Materiality of Traditional Ink Painting

实验水墨
绘画材料：宣纸，水墨

Experimental Ink Wash Painting
Artwork: Rice paper, Ink
21cm × 28cm
2023

实验水墨
绘画材料：素描纸，水墨
Experimental Ink Wash Painting
Artwork:Sketch paper, Ink
21cm×28cm
2023

实验水墨
绘画材料：素描纸，水墨
Experimental Ink Wash Painting
Artwork:Sketch paper, Ink
21cm×28cm
2023

实验水墨
绘画材料：素描纸，水墨

Experimental Ink Wash Painting
Artwork:Sketch paper, Ink
21cm × 28cm
2023

实验水墨
绘画材料：素描纸，水墨

Experimental Ink Wash Painting
Artwork: Sketch paper, Ink
21cm × 28cm
2023

实验水墨
绘画材料：速写纸，水墨

Experimental Ink Wash Painting
Artwork: Sketching paper, Ink
21cm×28cm
2023

实验水墨
绘画材料：素描纸，水墨

Experimental Ink Wash Painting
Artwork: Sketch paper, Ink
21cm × 28cm
2023

实验水墨
绘画材料：素描纸，水墨

Experimental Ink Wash Painting
Artwork:Sketch paper, Ink
21cm × 28cm
2023

实验水墨
绘画材料：宣纸，水墨

Experimental Ink Wash Painting
Artwork: Rice paper, Ink
33cm × 46cm
2023

实验水墨
绘画材料：宣纸板，水墨
Experimental Ink Wash Painting
Artwork：Rice paper board, Ink
33cm × 46cm
2023

实验水墨
绘画材料：素描纸，水墨
Experimental Ink Wash Painting
Artwork：Sketch paper, Ink
33cm × 46cm
2023

实验水墨
绘画材料：复印纸，水墨

Experimental Ink Wash Painting
Artwork: Duplicating paper, Ink
21cm × 28cm
2023

实验水墨
绘画材料：素描纸，水墨

Experimental Ink Wash Painting
Artwork:Sketch paper, Ink
21cm×28cm
2023

实验水墨
绘画材料：素描纸，水墨
Experimental Ink Wash Painting
Artwork: Sketch paper, Ink
21cm×28cm
2023

实验水墨
绘画材料：宣纸，水墨

Experimental Ink Wash Painting
Artwork: Rice paper, Ink
33cm × 46cm
2023

实验水墨
绘画材料：宣纸，水墨

Experimental Ink Wash Painting
Artwork: Rice paper, Ink
33cm × 46cm
2023

实验水墨
绘画材料：复印纸，水墨

Experimental Ink Wash Painting
Artwork: Duplicating paper, Ink
21cm×28cm
2023

实验水墨
绘画材料：素描纸，水墨

Experimental Ink Wash Painting
Artwork: Sketch paper, Ink
21cm×28cm
2023

"借由笔触皴擦点染的力道与章法，

我参与了水、墨、纸、笔质性的碰撞，

这一过程是充满乐趣与惊喜的。"

"Through the strokes, rubbings, and dyes

applied with the force and methods of brushwork,

I engage in the materialized collision of water, ink, paper, and the brush.

This process is filled with joy and surprises."

实验水墨
绘画材料：宣纸，水墨
Experimental Ink Wash Painting
Artwork:Rice paper, Ink
33cm×46cm
2023

实验水墨
绘画材料：宣纸，水墨

Experimental Ink Wash Painting
Artwork：Rice paper, Ink
33cm×46cm
2023

实验水墨
绘画材料：宣纸，水墨

Experimental Ink Wash Painting
Artwork：Rice paper, Ink
33cm×46cm
2023

实验水墨
绘画材料：宣纸，水墨

Experimental Ink Wash Painting
Artwork: Rice paper, Ink
33cm × 46cm
2023

实验水墨
绘画材料：宣纸，水墨

Experimental Ink Wash Painting
Artwork: Rice paper, Ink
33cm × 46cm
2023

实验水墨
绘画材料：宣纸，水墨
Experimental Ink Wash Painting
Artwork: Rice paper, Ink
33cm × 46cm
2023

实验水墨
绘画材料：宣纸，水墨
Experimental Ink Wash Painting
Artwork:Rice paper, Ink
33cm×46cm
2023

实验水墨
绘画材料：宣纸，水墨

Experimental Ink Wash Painting
Artwork：Rice paper, Ink
33cm × 46cm
2023

实验水墨
绘画材料：复印纸，水墨

Experimental Ink Wash Painting
Artwork: Duplicating paper, Ink
15cm × 16cm
2023

实验水墨
绘画材料：速写纸，水墨

Experimental Ink Wash Painting
Artwork：Sketching paper, Ink
15cm×16cm
2023

实验水墨
绘画材料：牛皮纸，水墨

Experimental Ink Wash Painting
Artwork：Kraft paper, Ink
21cm×28cm
2023

实验水墨
绘画材料：速写纸，水墨

Experimental Ink Wash Painting
Artwork: Sketching paper, Ink
21cm×28cm
2023

实验水墨
绘画材料：速写纸，水墨

Experimental Ink Wash Painting
Artwork:Sketching paper, Ink
21cm×28cm
2023

实验水墨
绘画材料：速写纸，水墨

Experimental Ink Wash Painting
Artwork:Sketching paper, Ink
21cm × 28cm
2023

实验水墨
绘画材料：速写纸，水墨

Experimental Ink Wash Painting
Artwork:Sketching paper, Ink
21cm × 28cm
2023

实验水墨
绘画材料：宣纸板，水墨

Experimental Ink Wash Painting
Artwork: Rice paper board, Ink
33cm × 46cm
2023

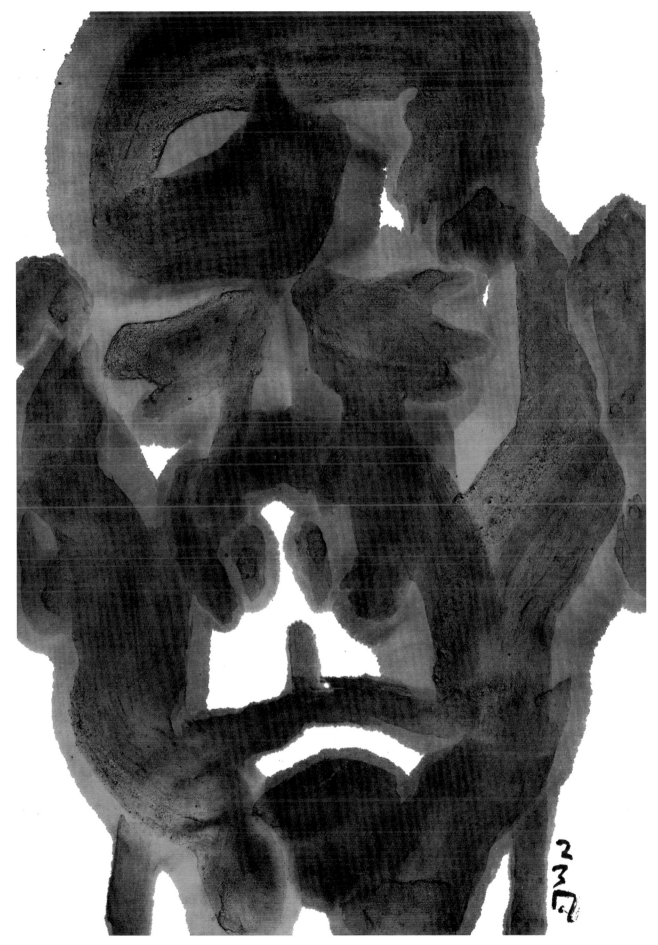

实验水墨
绘画材料：宣纸板，水墨

Experimental Ink Wash Painting
Artwork: Rice paper board, Ink
33cm × 46cm
2023

实验水墨
绘画材料：宣纸板，水墨

Experimental Ink Wash Painting
Artwork: Rice paper board, Ink
33cm × 46cm
2023

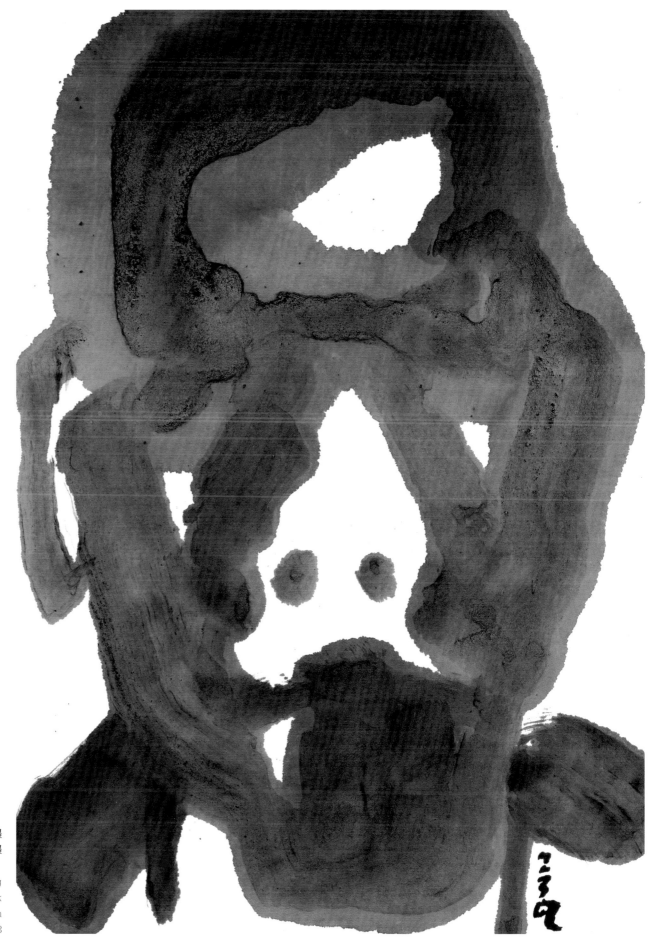

实验水墨
绘画材料：宣纸板，水墨
Experimental Ink Wash Painting
Artwork: Rice paper board, Ink
33cm × 46cm
2023

实验水墨
绘画材料：宣纸板，水墨

Experimental Ink Wash Painting
Artwork: Rice paper board, Ink
33cm × 46cm
2023

实验水墨
绘画材料：宣纸板，水墨

Experimental Ink Wash Painting
Artwork: Rice paper board, Ink
33cm × 46cm
2023

实验水墨
绘画材料：宣纸板，水墨

Experimental Ink Wash Painting
Artwork: Rice paper board, Ink
33cm × 46cm
2023

"我鲜少纠结于媒材的品质高低与媒材间的适配度，

而更注重表现

特定场景、特定人物所传达出的氛围、情绪等。"

"I don't get bogged down by the quality of the materials

or their conventional compatibility.

Instead, I focus more on expressing the atmosphere and emotions

conveyed by specific scenes and characters."

实验水墨
绘画材料：速写纸，水墨

Experimental Ink Wash Painting
Artwork：Sketching paper, Ink
21cm×28cm
2023

实验水墨
绘画材料：宣纸板，水墨
Experimental Ink Wash Painting
Artwork: Rice paper board, Ink
33cm × 46cm
2023

实验水墨
绘画材料：宣纸板，水墨

Experimental Ink Wash Painting
Artwork: Rice paper board, Ink
33cm × 46cm
2023

实验水墨
绘画材料：宣纸板，水墨
Experimental Ink Wash Painting
Artwork: Rice paper board, Ink
33cm × 46cm
2023

实验水墨
绘画材料：宣纸板，水墨
Experimental Ink Wash Painting
Artwork: Rice paper board, Ink
33cm×46cm
2023

实验水墨
绘画材料：宣纸板，水墨
Experimental Ink Wash Painting
Artwork: Rice paper board, Ink
33cm × 46cm
2023

实验水墨
绘画材料：宣纸板，水墨

Experimental Ink Wash Painting
Artwork: Rice paper board, Ink
33cm × 46cm
2023

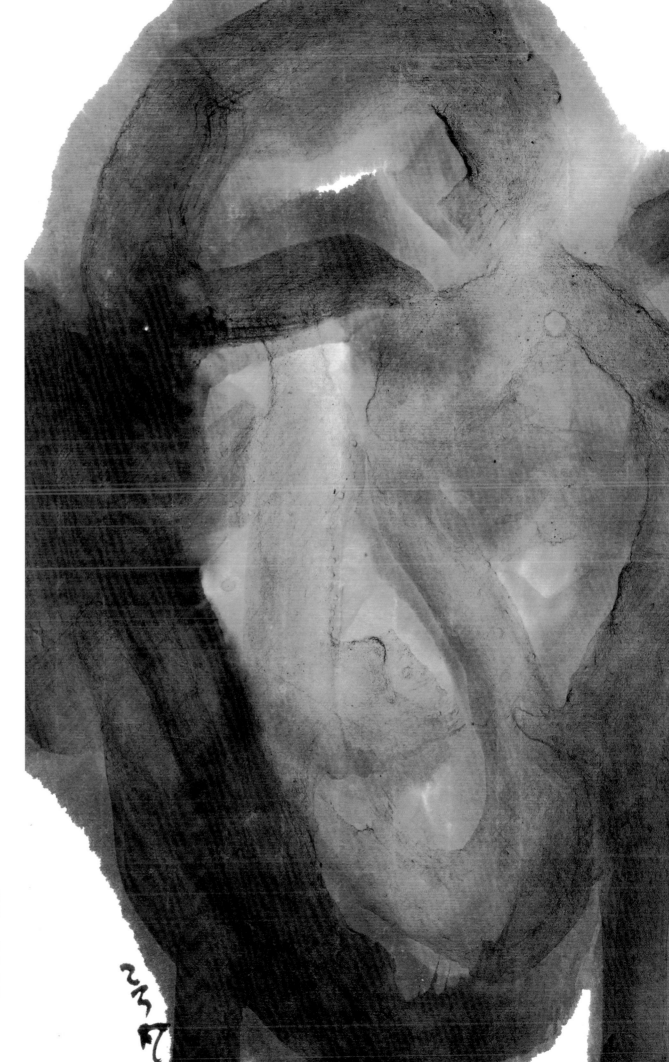

实验水墨
绘画材料：宣纸板，水墨

Experimental Ink Wash Painting
Artwork: Rice paper board, Ink
33cm × 46cm
2023

实验水墨
绘画材料：宣纸板，水墨

Experimental Ink Wash Painting
Artwork: Rice paper board, Ink
33cm × 46cm
2023

实验水墨
绘画材料：宣纸板，水墨
Experimental Ink Wash Painting
Artwork: Rice paper board, Ink
33cm×46cm
2023

实验水墨
绘画材料：宣纸板，水墨

Experimental Ink Wash Painting
Artwork: Rice paper board, Ink
33cm × 46cm
2023

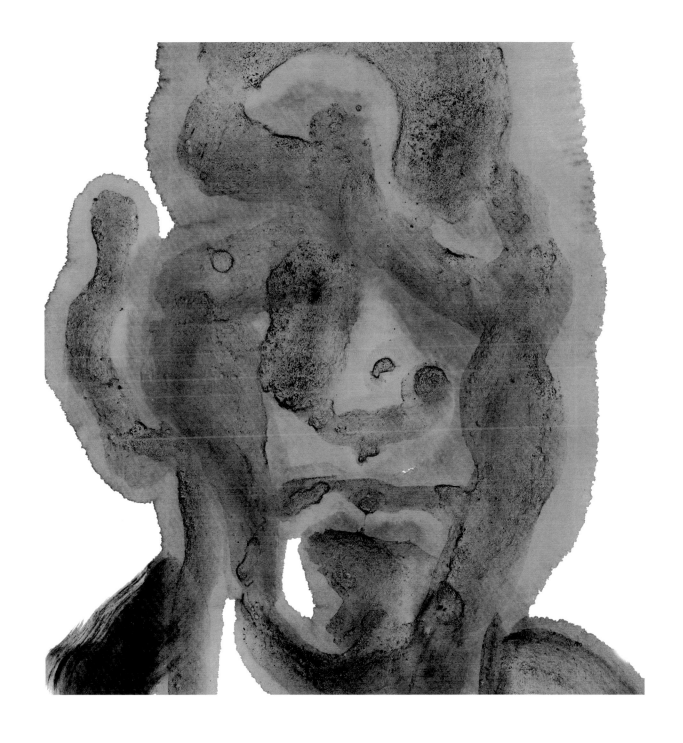

实验水墨
绘画材料：宣纸板，水墨

Experimental Ink Wash Painting
Artwork: Rice paper board, Ink
33cm×46cm
2023

实验水墨
绘画材料：牛皮纸，水墨

Experimental Ink Wash Painting
Artwork：Kraft paper, Ink
21cm×28cm
2023

实验水墨
绘画材料：牛皮纸，水墨

Experimental Ink Wash Painting
Artwork：Kraft paper, Ink
21cm×28cm
2023

实验水墨
绘画材料：牛皮纸，水墨

Experimental Ink Wash Painting
Artwork: Kraft paper, Ink
55cm × 60cm
2023

实验水墨
绘画材料：牛皮纸，水墨

Experimental Ink Wash Painting
Artwork:Kraft paper, Ink
55cm×60cm
2023

实验水墨
绘画材料:牛皮纸,水墨

Experimental Ink Wash Painting
Artwork:Kraft paper, Ink
55cm×60cm
2023

实验水墨
绘画材料：牛皮纸，水墨

Experimental Ink Wash Painting
Artwork：Kraft paper, Ink
33cm × 46cm
2023

"对于艺术创作，理性固然重要，

但往往是不期而遇的、不确定的感性

才是能够激活一幅画作艺术灵韵的存在。"

"For artistic creation, rationality is undoubtedly important,

but it is the unexpected and uncertain sensibility

that can activate the artistic charm of a painting."

实验水墨
绘画材料：牛皮纸，水墨

Experimental Ink Wash Painting
Artwork: Kraft paper, Ink
55cm × 60cm
2023

实验水墨
绘画材料：信封纸，水墨
Experimental Ink Wash Painting
Artwork: Envelope paper, Ink
55cm×60cm
2023

实验水墨
绘画材料：牛皮纸，水墨

Experimental Ink Wash Painting
Artwork: Kraft paper, Ink
55cm × 60cm
2023

实验水墨
绘画材料：牛皮纸，水墨
Experimental Ink Wash Painting
Artwork: Kraft paper, Ink
55cm×60cm
2023

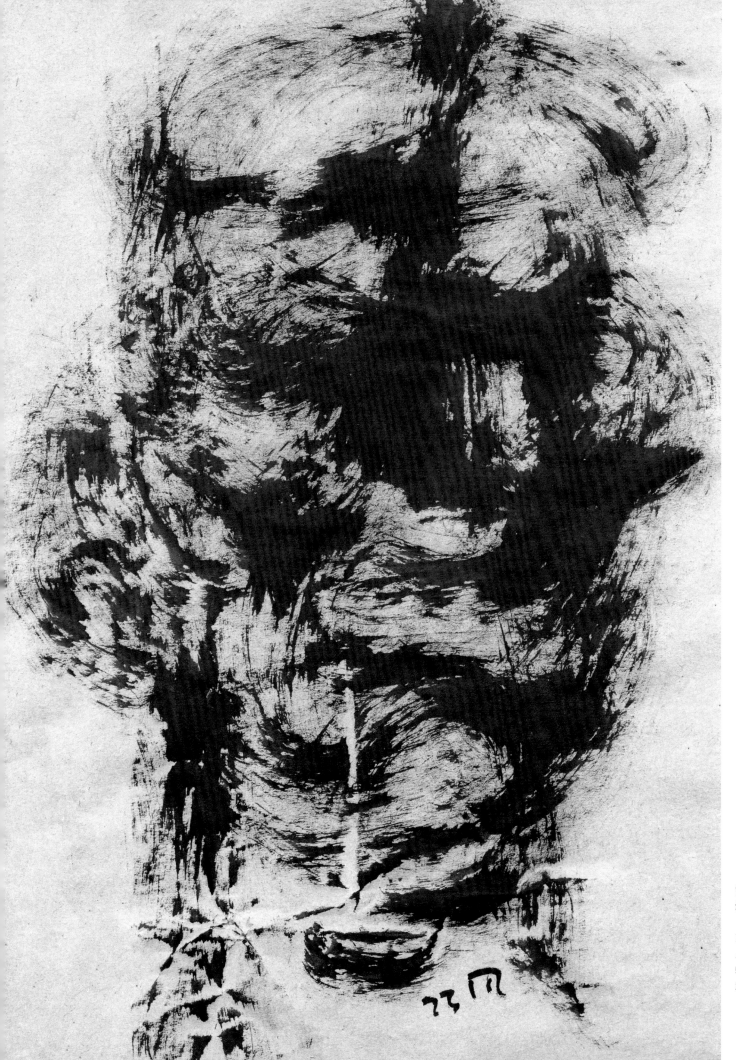

实验水墨
绘画材料：牛皮纸，水墨

Experimental Ink Wash Painting
Artwork:Kraft paper, Ink
55cm×60cm
2023

实验水墨
绘画材料：牛皮纸，水墨
Experimental Ink Wash Painting
Artwork: Kraft paper, Ink
55cm × 60cm
2023

实验水墨
绘画材料：复印纸，水墨

Experimental Ink Wash Painting
Artwork:Duplicating paper, Ink
21cm × 28cm
2023

实验水墨
绘画材料：信封纸，水墨

Experimental Ink Wash Painting
Artwork：Envelope paper, Ink
21cm × 28cm
2023

实验水墨
绘画材料：宣纸，水墨

Experimental Ink Wash Painting
Artwork: Rice paper, Ink
21cm×28cm
2023

实验水墨
绘画材料：宣纸，水墨

Experimental Ink Wash Painting
Artwork: Rice paper, Ink
21cm×28cm
2023

实验水墨
绘画材料：速写纸，水墨

Experimental Ink Wash Painting
Artwork:Sketching paper, Ink
21cm×28cm
2023

实验水墨
绘画材料：素描纸，水墨

Experimental Ink Wash Painting
Artwork: Sketch paper, Ink
21cm × 28cm
2023

实验水墨
绘画材料：宣纸，水墨

Experimental Ink Wash Painting
Artwork: Rice paper, Ink
21cm × 28cm
2023

第二章
临场：见证水墨艺术未知的多样性

Chapter 2
On Scene: Witnessing the Unknown of Ink Art

采访04：上次咱们讨论了传统水墨绘画的质料性，您认为"无纸化"的创作方式对于水墨绘画而言意味着什么？

凡： 这个问题需要兼顾画作与画家两个维度进行探讨。从画作的维度看，现今我们能通过Procreate等电绘软件，模仿毛笔、水彩、铅笔、炭笔等笔触，甚至可以实现墨在水中扩散、消解的过程以及墨在不同材质的纸张上，配合不同的笔法晕染开来的效果。与传统意义上美术馆中经装裱、壁挂、布光等环节后所展陈的画作不同，"无纸化"模式下创作的画作可以让观者仅凭双指触屏放大便可清晰品鉴作品的局部，可以说画作与观者的距离被无限拉近，这对画面细节的雕琢精度提出了更高的要求。配合数字影像技术的应用，我们甚至可以让静态的水墨形象动起来，以动画的形式呈现水墨神韵，使古典的水墨艺术在当代技术语境下得到更为丰富、多元的艺术表现形式，这也是我在个人执导的8K水墨动画短片《秋实》《立秋》中所试图呈现的。

从画家的维度看，我认为"无纸化"的创作方式对习惯了笔墨纸砚的传统中国画艺术家而言无疑是具有高度挑战性的——他们需要重新适应一套新的创作模式，在技术应用层面上，需要通过不断的练习以适应如手写板上压感笔的笔触调节等；在艺术形式层面上，需要不断关注技术发展的前沿态势，思考作为古典艺术形式的水墨如何在当代技术视阈下实现自身的创造性转化与创新性发展；进而在艺术观念层面，思考水墨于当代人文生态下可表达的作品题材与精神内核是否有转向或外延……这些都是创作者需要时刻考虑的关键问题。

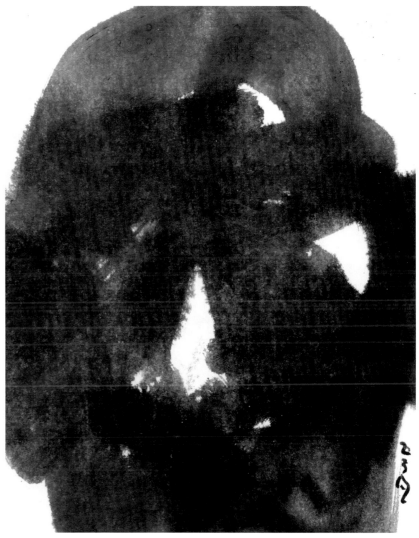

实验水墨
绘画材料：特种纸，水墨

Experimental Ink Wash Painting
Artwork:Specialty paper, Ink
21cm×28cm
2023

实验水墨
绘画材料：特种纸，水墨

Experimental Ink Wash Painting
Artwork:Specialty paper, Ink
21cm×28cm
2023

Q4: We have discussed the materiality of traditional ink painting. What kind of impact do you think the "paperless" creation method has on ink painting?

Fan: This question requires consideration from both the perspective of the artwork and the artist. For the artwork, nowadays, with digital drawing software like Procreate, we can mimic brushstrokes such as those of a calligraphy brush, watercolor brush, pencil, charcoal, and even simulate the diffusion and dissolution of ink in water, as well as the effects of ink spreading and blending on different types of paper with various drawing techniques. Unlike artworks traditionally displayed in museums after being framed, hung, and lit from certain angles, artworks created in a "paperless" mode allow viewers to zoom in to get a closer look at specific parts of the artwork with just two fingers on a touchscreen, bringing the artwork closer to the viewer infinitely. This increases the demand for precision in refining the details of the image. With the application of digital imaging technology, we can even animate static ink images, presenting the charm of ink painting in the form of animation. This enriches the classical ink art with more diverse artistic expressions in the context of contemporary technology. This is also what I attempted to present in the 8K ink animation short films *Qiu Shi* and *Li Qiu*.

From the POV of the artist, I believe the "paperless" creation method is undoubtedly highly challenging for traditional Chinese ink painting artists accustomed to brush, ink, paper, and inkstone—it requires them to adapt to a new set of creative processes. On the technical level, they need to practice continuously to adapt to features such as pressure-sensitive pens on digital tablets. On the artistic level, they need to pay attention to the forefront of technological development, contemplate how ink, as a classical art form, can achieve creative transformation and innovative development in the context of contemporary technology. Furthermore, on the level of artistic concepts, they need to consider whether there is a shift or extension in the themes and spiritual value that ink painting can express within contemporary human and ecological contexts. These are all crucial issues that creators need to constantly consider.

Chapter 2　On Scene: Witnessing the Unknown of Ink Art

实验水墨
绘画材料：宣纸，水墨、水彩

Experimental Ink Wash Painting
Artwork: Rice paper, Ink, Watercolor
55cm × 60cm
2023

采访 05：您认为探索水墨艺术在当代技术视阈下的多元形态，对水墨艺术本身而言的现实意义为何？

凡： 艺术史学家贡布里希说过："以各种各样的观念和媒材来从事实验的自由是 20 世纪艺术的特征。"对于水墨而言亦是如此，当代技术的发展赋予水墨艺术以开放的表现形式，极大程度上扩展了传统意义上"水墨"的外延，实现了现代化的形象思维、符号图示乃至精神内核与水墨"浓淡虚实""气韵生动"的艺术风格间的契合。对于水墨艺术而言，其现实意义首先是激励了新生代创作者不断继承传统，推陈出新，将水墨由绘画艺术延展至影像、装置、雕塑等多元艺术形态中，基于"已有"创作"未有"，让观者在新奇的观看方式、丰富的视觉图式与多元的表现手法中更为直观且沉浸地感受水墨材质性能的肌理之美，并在此基础上试图与当代国人的心理诉求产生一定的呼应，彰显当代国人火热的生活图景，让水墨艺术成为一种彰显艺术家从审美与心灵层面体察自我、他人与社会动态关系的载体，将水墨艺术实验与生活经验、社会体验、情感超验串联于一处，追求"道术相融""气韵生动"的美学风格，让当代中国水墨艺术能够以民族性、时代性、创新性走向世界。

实验水墨
绘画材料：宣纸板，水墨

Experimental Ink Wash Painting
Artwork: Rice paper board, Ink
33cm × 46cm
2023

Q5: What do you think is the practical significance of exploring diverse forms of ink painting with the application of different kinds of contemporary technology for ink painting itself?

Fan: The art historian Gombrich once said, "The freedom to experiment with various ideas and media is a characteristic of 20th-century art". The same applies to ink painting. The development of contemporary technology has endowed ink painting with open forms of expression, greatly expanding the traditional understanding of "ink". This has realized a fusion between modern image thinking method, symbolic representation, and the spiritual expression of ink painting, such as the degree of shadings and reality, the charismatic of charm and vitality etc. For ink painting, its significance lies firstly in inspiring new-generation creators to continuously inherit tradition, based on the heritage we innovate, and extend ink painting from the realm of painting into various art forms such as imagery, installations, and sculptures. They base their creations on what already exists while creating what is yet to come. This allows viewers to more intuitively and deeply experience the beauty of the texture of ink materials through surprising ways of viewing, rich visual patterns, and diverse expressive techniques. Furthermore, it attempts to resonate with the psychological aspirations of contemporary Chinese people, showcasing the vibrant scenes of their daily life. Ink painting thus becomes a carrier for artists to observe their self, others, and the dynamic relationship with society from aesthetic and spiritual perspectives. It connects ink painting experiments with

Chapter 2　On Scene: Witnessing the Unknown of Ink Art

life experiences, social experiences, and transcendent emotions, seeking an aesthetic style of blending classical Daoism with vivid vitality. This allows contemporary Chinese ink painting to embrace national, contemporary, and innovative qualities as it ventures its way worldwide.

写生习作
绘画材料：复印纸，炭笔

Life Drawing Practice Painting
Artwork:Duplicating paper,Charcoal
21cm×28cm
2023

第二章　临场：见证水墨艺术未知的多样性

采访06：您认为未来水墨艺术将如何实现多元的发展？

凡： 这一点是我在个人创作中不断探索的。我想所谓的"多元"，首先要跳脱出"画"本身，通过关注"画"的内涵与外延来寻求多元的发展路径。比如在本书中收录了一些相机拍摄的图像，我将其称之为水墨绘画的"遗址"，所谓遗址的特性，首先表现为"不完整的残存物"，其次还特指具备"人类活动"的遗迹，从史学、美学、人类学等角度看具有突出的普遍价值。在这些被称作是"遗址"的图像中不仅可以看到画作的成品，还可以看到画作旁侧的茶杯、晕开笔墨的杯盖、积灰的砚台、斑驳的调色盘等等，比起陈列在美术馆里装裱后打上光的作品，这种图片看似不够考究，但它得以呈现画者自身在创作当下所体验的"临场感"，于我而言更是生动地留存了"即兴"发生的当下所身处的环境，一定程度上有助于观者将视线从"画"本身延展至"绘画的人""绘画的当下"等维度，进而丰富对画作"从无到有"过程性的理解。这是我个人所做出的一点展陈形式方面的尝试，未来水墨艺术或将在作品题材、创作手法、工艺应用等多个维度不断突破自身的"舒适圈"。如水与墨互动的形态千变万化、难以复刻一般，我相信未来水墨艺术的形态也将伴随着技术的演进而不断丰富。

Chapter 2 On Scene: Witnessing the Unknown of Ink Art

实验水墨
绘画材料：宣纸，水墨、水彩
Experimental Ink Wash Painting
Artwork:Rice paper, Ink, Watercolor
55cm×60cm
2023

第二章　临场：见证水墨艺术未知的多样性

Q6: How can ink painting achieve diverse development in the future?

Fan: This is something I have been continuously exploring in my personal practices. I believe that the so-called "diversity" first needs to break free from the concept of "painting" itself and instead focus on the connotation and extension of "painting" to seek diverse possibilities that is yet to come. For example, in this book, I have included some images taken by a camera, which I refer to as the "ruins" of ink painting. The characteristics of these "ruins" firstly manifest as "incomplete remnants", and secondly, they specifically refer to traces left by human activities, which possess significant universal value historically, aesthetically and more. In these images labeled as "ruins", one can not only see the finished artworks but also tea cups beside the paintings, cup lids where brushes are soaked, dusty inkstones, and mottled palettes. Compared to meticulously displayed artworks framed and illuminated in museums, these images may seem less refined. However, they lively capture the "sense of on-scene" experienced by the artist during creation. For me, they also vividly preserve the environment in which spontaneity occurs, to some extent assisting viewers in expanding their focus from the artwork itself to dimensions such as the "painter" and the "very moment of drawing", thereby enriching their understanding of the process from nothing to something that occurred during the creation of the artwork. This is one form of exhibition attempt I have made personally. In the future, ink painting may continuously break through its own "comfort zone" in terms of subject matter, creative techniques, and craft applications. Just as the forms of interaction between water and ink are ever-changing and difficult to replicate, I believe the forms of ink painting in the future will also continuously enrich themselves along with the evolution of technology.

Chapter 2 On Scene: Witnessing the Unknown of Ink Art

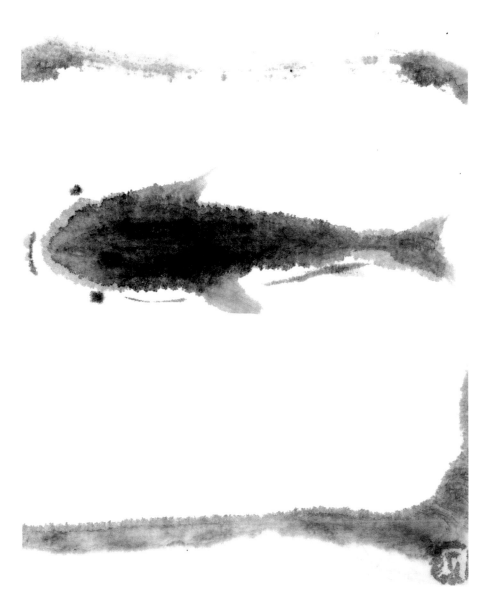

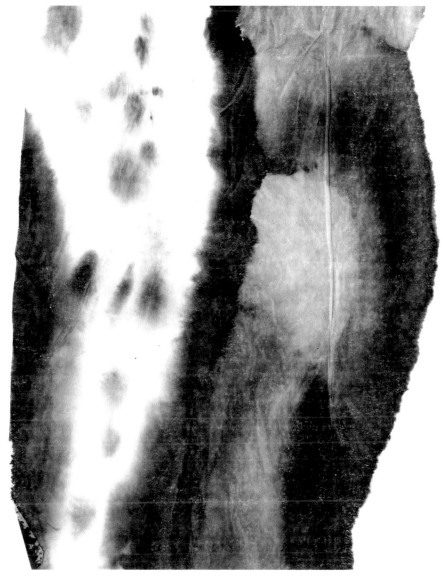

实验水墨
绘画材料：餐巾纸，水墨，茶水
Experimental Ink Wash Painting
Artwork:Napkin paper, Ink, Tea
12cm × 14cm
2023

实验水墨
绘画材料：餐巾纸，水墨
Experimental Ink Wash Painting
Artwork:Napkin paper, Ink
12cm × 14cm
2023

实验水墨
绘画材料：宣纸，水墨

Experimental Ink Wash Painting
Artwork: Rice paper, Ink
21cm × 28cm
2023

实验水墨
绘画材料：宣纸，水墨

Experimental Ink Wash Painting
Artwork: Rice paper, Ink
21cm × 28cm
2023

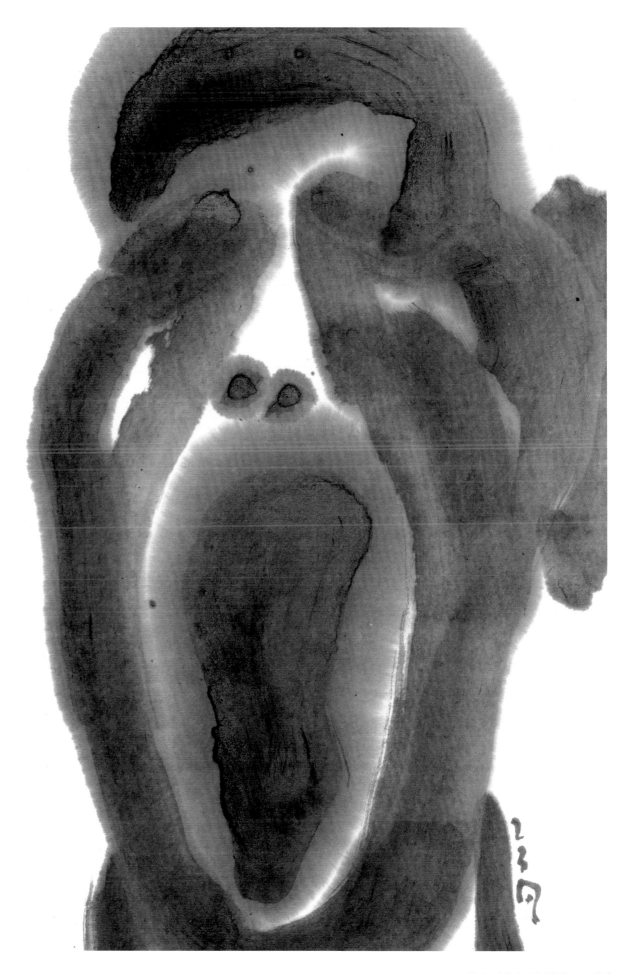

实验水墨
绘画材料：宣纸，水墨

Experimental Ink Wash Painting
Artwork: Rice paper, Ink
21cm × 28cm
2023

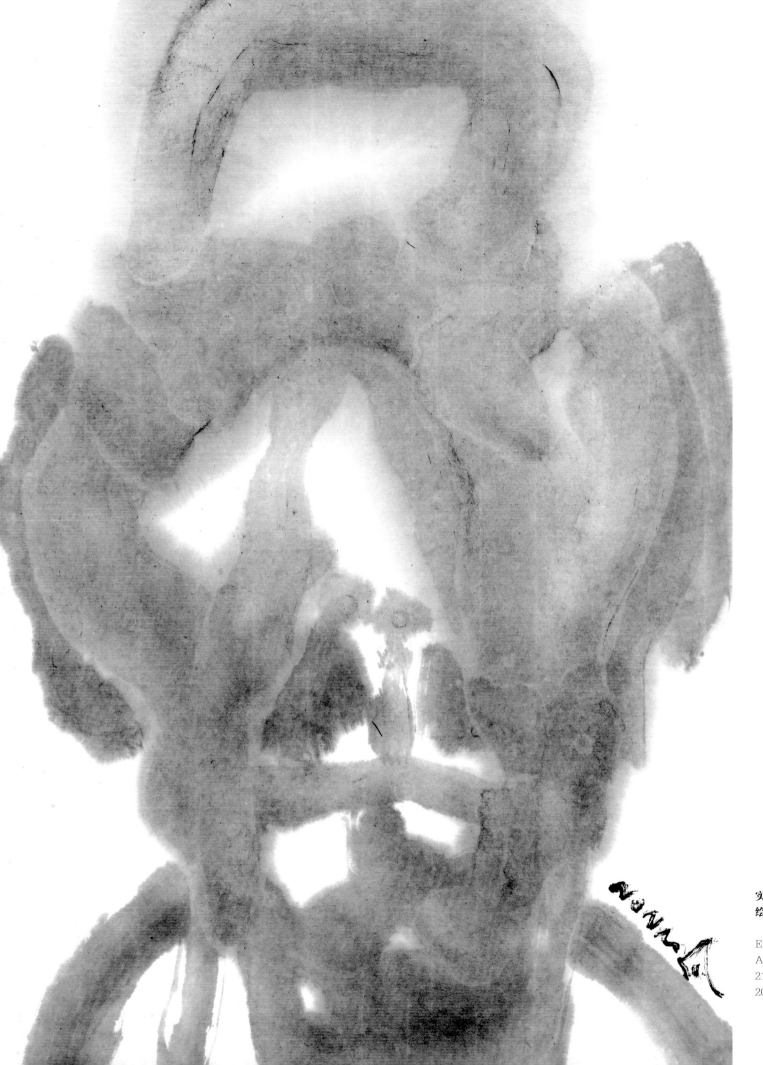

实验水墨
绘画材料：宣纸，水墨

Experimental Ink Wash Painting
Artwork: Rice paper, Ink
21cm×28cm
2023

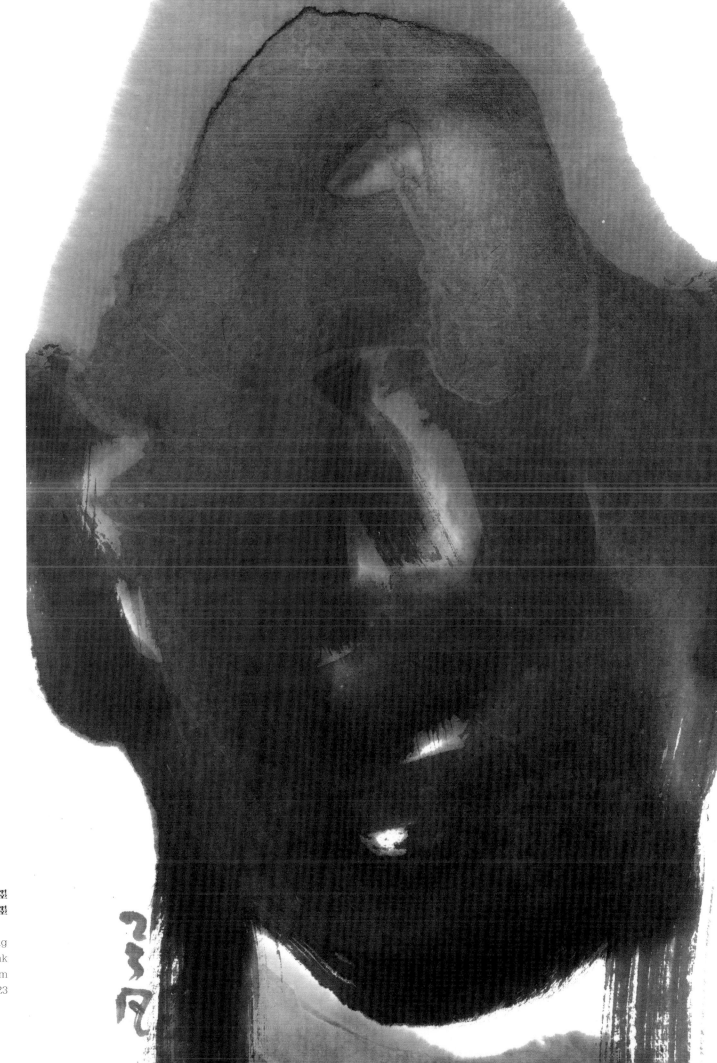

实验水墨
绘画材料：宣纸，水墨

Experimental Ink Wash Painting
Artwork:Rice paper, Ink
21cm×28cm
2023

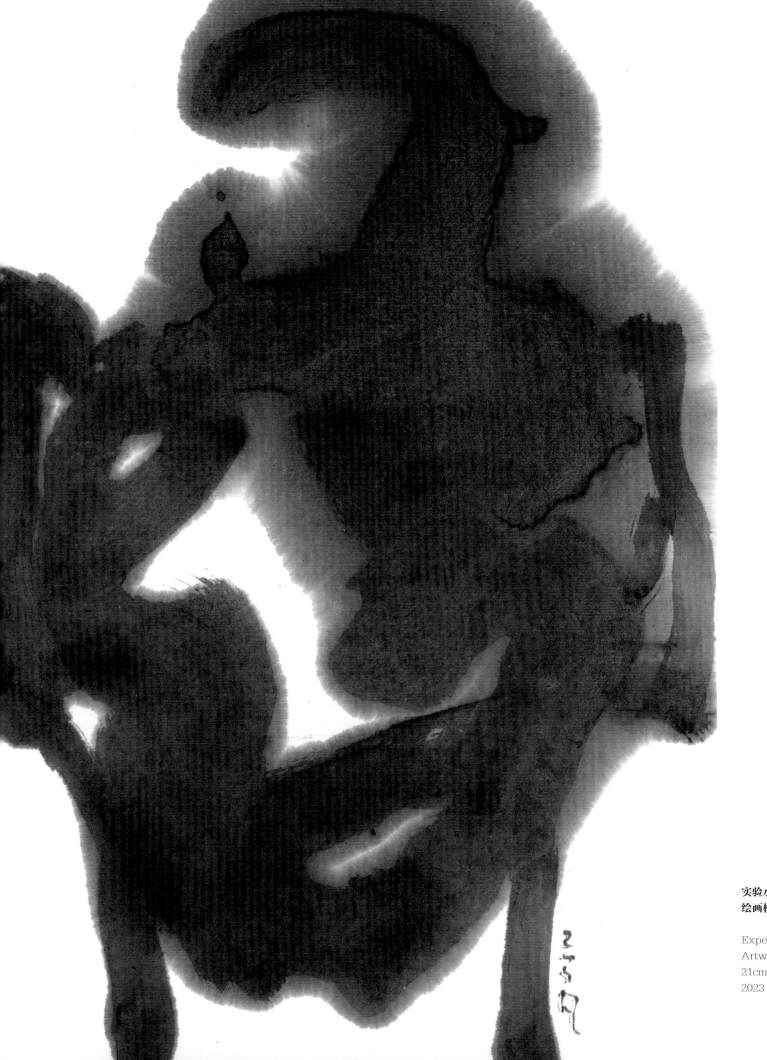

实验水墨
绘画材料：宣纸，水墨

Experimental Ink Wash Painting
Artwork: Rice paper, Ink
21cm × 28cm
2023

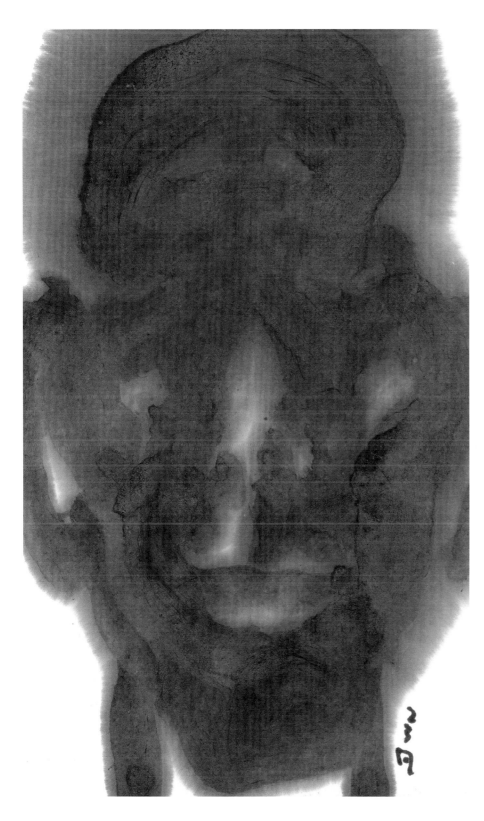

实验水墨
绘画材料：宣纸，水墨

Experimental Ink Wash Painting
Artwork：Rice paper, Ink
21cm×28cm
2023

实验水墨
绘画材料：宣纸，水墨

Experimental Ink Wash Painting
Artwork：Rice paper, Ink
21cm×28cm
2023

实验水墨
绘画材料：宣纸，水墨

Experimental Ink Wash Painting
Artwork:Rice paper, Ink
21cm×28cm
2023

实验水墨
绘画材料：宣纸，水墨

Experimental Ink Wash Painting
Artwork: Rice paper, Ink
21cm × 28cm
2024

实验水墨
绘画材料：复印纸，水墨

Experimental Ink Wash Painting
Artwork: Duplicating paper, Ink
12cm×14cm
2023

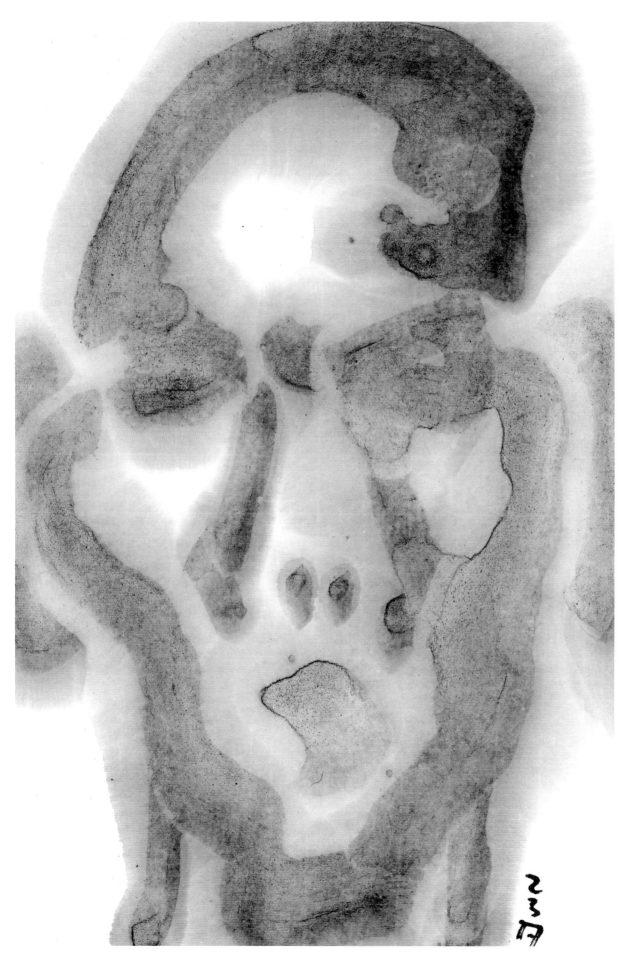

实验水墨
绘画材料：宣纸板，水墨

Experimental Ink Wash Painting
Artwork: Rice paper board, Ink
33cm×46cm
2023

实验水墨
绘画材料：素描纸，水墨

Experimental Ink Wash Painting
Artwork:Sketch paper, Ink
33cm×46cm
2023

实验水墨
绘画材料：宣纸，水墨

Experimental Ink Wash Painting
Artwork: Rice paper, Ink
55cm × 60cm
2023

实验水墨
绘画材料：宣纸，水墨

Experimental Ink Wash Painting
Artwork: Rice paper, Ink
55cm×60cm
2022

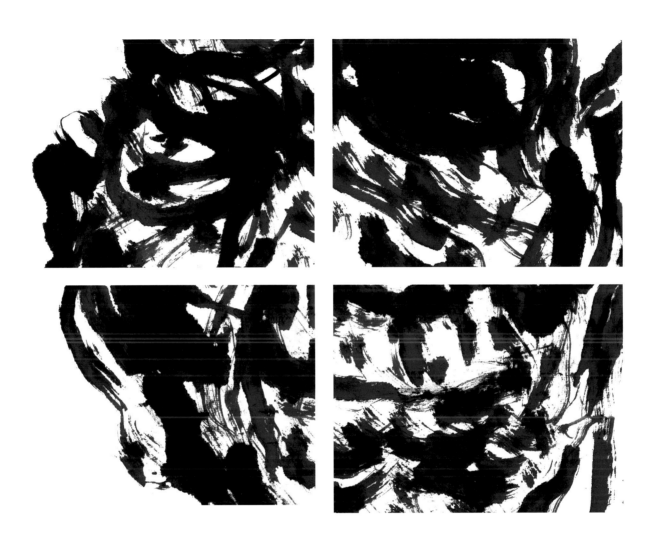

实验水墨
绘画材料：素描纸，水墨

Experimental Ink Wash Painting
Artwork:Sketch paper, Ink
10cm × 16cm
2023

实验水墨
绘画材料：素描纸，水墨

Experimental Ink Wash Painting
Artwork: Sketch paper, Ink
10cm × 16cm
2023

"未来水墨艺术当基于'已有'创作'未有',

让观者在新奇的观看方式、丰富的视觉图式与多元的表现手法中更为直观且沉浸地感受

水墨材质的肌理之美,

并在此基础上试图与当代国人的心理诉求产生一定的呼应。"

"They base their creations on what already exists while creating what is yet to come.

This allows viewers to more intuitively and deeply experience

the beauty of the texture of ink materials through surprising ways of viewing,

rich visual patterns, and diverse expressive techniques.

then makes it possible to resonate with the psychological aspirations of contemporary

Chinese people."

实验水墨
绘画材料：速写纸，水墨

Experimental Ink Wash Painting
Artwork:Sketching paper, Ink
10cm×16cm
2023

实验水墨
绘画材料：速写纸，水墨
Experimental Ink Wash Painting
Artwork:Sketching paper, Ink
10cm×16cm
2023

实验水墨
绘画材料：速写纸，水墨

Experimental Ink Wash Painting
Artwork:Sketching paper, Ink
10cm × 16cm
2023

实验水墨
绘画材料：速写纸，水墨

Experimental Ink Wash Painting
Artwork:Sketching paper, Ink
10cm×16cm
2023

实验水墨
绘画材料：速写纸，水墨

Experimental Ink Wash Painting
Artwork：Sketching paper, Ink
10cm×16cm
2023

实验水墨
绘画材料：速写纸，水墨
Experimental Ink Wash Painting
Artwork: Sketching paper, Ink
10cm × 16cm
2023

实验水墨
绘画材料：速写纸，水墨
Experimental Ink Wash Painting
Artwork：Sketching paper, Ink
10cm × 16cm
2023

实验水墨
绘画材料：速写纸，水墨

Experimental Ink Wash Painting
Artwork:Sketching paper, Ink
10cm×16cm
2023

实验水墨
绘画材料：速写纸，水墨

Experimental Ink Wash Painting
Artwork:Sketching paper, Ink
10cm × 16cm
2023

实验水墨
绘画材料:速写纸,水墨
Experimental Ink Wash Painting
Artwork:Sketching paper, Ink
10cm × 16cm
2023

实验水墨
绘画材料：宣纸，水墨

Experimental Ink Wash Painting
Artwork：Rice paper, Ink
55cm×60cm
2023

实验水墨
绘画材料：牛皮纸，水墨

Experimental Ink Wash Painting
Artwork：Kraft paper, Ink
55cm×60cm
2023

实验水墨
绘画材料：复印纸，水墨

Experimental Ink Wash Painting
Artwork: Duplicating paper, Ink
21cm × 28cm
2023

实验水墨
绘画材料：牛皮纸，炭笔
Experimental Ink Wash Painting
Artwork：Kraft paper, Charcoal
55cm×60cm
2023

实验水墨
绘画材料：特种纸，水墨

Experimental Ink Wash Painting
Artwork:Specialty paper, Ink
21cm×28cm
2023

实验水墨
绘画材料：特种纸，水墨

Experimental Ink Wash Painting
Artwork:Specialty paper, Ink
21cm×28cm
2023

实验水墨
绘画材料：复印纸，水墨
Experimental Ink Wash Painting
Artwork: Duplicating paper, Ink
33cm × 46cm
2024

实验水墨
绘画材料：特种纸，水墨

Experimental Ink Wash Painting
Artwork:Specialty paper, Ink
33cm×46cm
2023

"将水墨艺术实验与生活经验、社会体验、情感超验串联于一处,

追求'道术相融''气韵生动'的美学风格,

让当代中国水墨艺术能够以民族性、时代性、创新性走向世界。"

"By connecting ink painting experiments with

life experiences, social experiences, and transcendent emotions together as one,

we are seeking an aesthetic style that blends classical Daoism with vivid vitality.

This allows contemporary Chinese ink painting to embrace

national, contemporary, and innovative qualities as it ventures its way worldwide."

实验水墨
绘画材料：宣纸板，水墨

Experimental Ink Wash Painting
Artwork：Rice paper board, Ink
33cm×46cm
2023

实验水墨
绘画材料：宣纸板，水墨

Experimental Ink Wash Painting
Artwork：Rice paper board, Ink
33cm×46cm
2023

实验水墨
绘画材料：宣纸板，水墨

Experimental Ink Wash Painting
Artwork: Rice paper board, Ink
33cm × 46cm
2023

实验水墨
绘画材料：宣纸板，水墨
Experimental Ink Wash Painting
Artwork: Rice paper board, Ink
33cm×46cm
2023

实验水墨
绘画材料：素描纸，水墨

Experimental Ink Wash Painting
Artwork：Sketch paper, Ink
21cm × 28cm
2023

基于左图的 AI 衍生数字绘画
媒材模拟：素描纸，水墨

AI-derived digital painting based on the painting (Left)
Simulated Material: Sketch paper, Ink
1856px × 2646px
2023

实验水墨
绘画材料：素描纸，水墨

Experimental Ink Wash Painting
Artwork: Sketch paper, Ink
33cm × 46cm
2023

基于左图的 AI 衍生数字绘画
媒材模拟：素描纸，水墨

AI-derived digital painting based on the painting (Left)
Simulated Material: Sketch paper, Ink
1856px × 2464px
2023

实验水墨	基于左图的 AI 衍生数字绘画
绘画材料：宣纸，水墨	媒材模拟：宣纸，水墨
Experimental Ink Wash Painting	AI-derived digital painting based on the painting (Left)
Artwork：Rice paper, Ink	Simulated Material：Rice paper, Ink
33cm × 46cm	928px × 1232px
2023	2023

实验水墨
绘画材料：宣纸，水墨

Experimental Ink Wash Painting
Artwork：Rice paper, Ink
33cm×46cm
2023

基于左图的 AI 衍生数字绘画
媒材模拟：宣纸，水墨

AI-derived digital painting based on the painting (Left)
Simulated Material：Rice paper, Ink
1856px×2464px
2023

实验水墨 基于左图的 AI 衍生数字绘画
绘画材料：素描纸，水墨 媒材模拟：素描纸，水墨

Experimental Ink Wash Painting AI-derived digital painting based on the painting (Left)
Artwork：Sketch paper, Ink Simulated Material：Sketch paper, Ink
33cm×46cm 928px×1232px
2023 2023

基于右图的 AI 衍生数字绘画 　　实验水墨
媒材模拟：宣纸，水墨 　　绘画材料：宣纸，水墨

AI-derived digital painting based on the painting (Right)　　Experimental Ink Wash Painting
Simulated Material：Rice paper, Ink　　Artwork：Rice paper, Ink
928px × 1232px　　33cm × 46cm
2023　　2023

实验水墨
绘画材料：素描纸，水墨

Experimental Ink Wash Painting
Artwork：Sketch paper, Ink
33cm×46cm
2023

基于上图的 AI 衍生数字绘画
媒材模拟：素描纸，水墨

AI-derived digital painting based on the painting (Upper)
Simulated Material：Sketch paper, Ink
1856px×2464px
2023

基于右图的 AI 衍生数字绘画　　实验水墨
媒材模拟：宣纸，水墨　　绘画材料：宣纸，水墨

AI-derived digital painting based on the painting (Right)　　Experimental Ink Wash Painting
Simulated Material:Rice paper, Ink　　Artwork:Rice paper, Ink
1856px × 2464px　　33cm × 46cm
2023　　2023

基于右图的 AI 衍生数字绘画　　实验水墨
媒材模拟：宣纸，水墨　　　　　绘画材料：宣纸，水墨

AI-derived digital painting based on the painting (Right)　　Experimental Ink Wash Painting
Simulated Material:Rice paper, Ink　　　　　　　　　　　　Artwork:Rice paper, Ink
1856px × 2464px　　　　　　　　　　　　　　　　　　　　33cm × 46cm
2023　　　　　　　　　　　　　　　　　　　　　　　　　2023

实验水墨
绘画材料：宣纸，水墨

Experimental Ink Wash Painting
Artwork : Rice paper, Ink
33cm × 46cm
2023

基于左图的 AI 衍生数字绘画
媒材模拟：宣纸，水墨

AI-derived digital painting based on the painting (Left)
Simulated Material : Rice paper, Ink
1856px × 2464px
2023

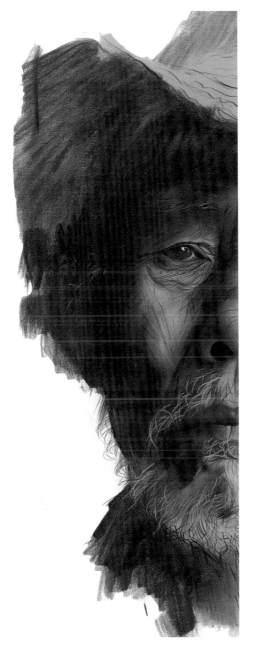 

基于右图的 AI 衍生数字绘画　　实验水墨
媒材模拟：牛皮纸，水墨　　绘画材料：牛皮纸，水墨

AI-derived digital painting based on the painting (Right)　　Experimental Ink Wash Painting
Simulated Material: Kraft paper, Ink　　Artwork: Kraft paper, Ink
1856px × 2464px　　33cm × 46cm
2023　　2023

实验水墨
绘画材料：牛皮纸，水墨

Experimental Ink Wash Painting
Artwork：Kraft paper, Ink
33cm×46cm
2023

基于上图的 AI 衍生数字绘画
媒材模拟：牛皮纸，水墨

AI-derived digital painting based on the painting (Upper)
Simulated Material：Kraft paper, Ink
1856px×2464px
2023

实验水墨 基于左图的 AI 衍生数字绘画
绘画材料：速写纸，水墨 媒材模拟：速写纸，水墨

Experimental Ink Wash Painting AI-derived digital painting based on the painting (Left)
Artwork：Sketching paper, Ink Simulated Material：Sketching paper, Ink
21cm×28cm 1856px×2464px
2023 2023

"我想所谓的'多元',

首先要跳脱出'画'本身,

通过关注'画'的内涵与外延来寻求多元的发展路径。"

"I believe that the so-called 'diversity' first

needs to break free from the concept of 'painting' itself

and instead focus on the connotation

and extension of 'painting' to seek diverse possibilities that is yet to come."

实验水墨 　　　　　　　　　基于左图的 AI 衍生数字绘画
绘画材料：牛皮纸，水墨　　　媒材模拟：牛皮纸，水墨

Experimental Ink Wash Painting　　AI-derived digital painting based on the painting (Left)
Artwork：Kraft paper, Ink　　　　Simulated Material：Kraft paper, Ink
55cm×60cm　　　　　　　　　928px×1232px
2023　　　　　　　　　　　　2023

实验水墨
绘画材料：牛皮纸，水墨
Experimental Ink Wash Painting
Artwork：Kraft paper, Ink
55cm×60cm
2023

基于左图的 AI 衍生数字绘画
媒材模拟：牛皮纸，水墨
AI-derived digital painting based on the painting (Left)
Simulated Material：Kraft paper, Ink
1856px×2464px
2023

实验水墨
绘画材料：速写纸，水墨

Experimental Ink Wash Painting
Artwork：Sketching paper, Ink
21cm×28cm
2023

基于左图的 AI 衍生数字绘画
媒材模拟：速写纸，水墨

AI-derived digital painting based on the painting (Left)
Simulated Material: Sketching paper, Ink
2048px × 2048px
2023

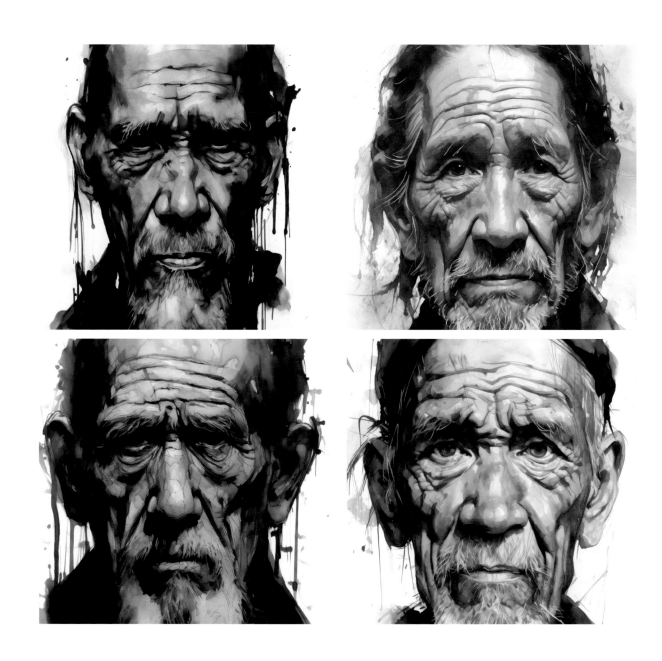

基于右图的 AI 衍生数字绘画
媒材模拟：宣纸，水墨

AI-derived digital painting based on the painting (Right)
Simulated Material: Rice paper, Ink
1928px × 2942px
2023

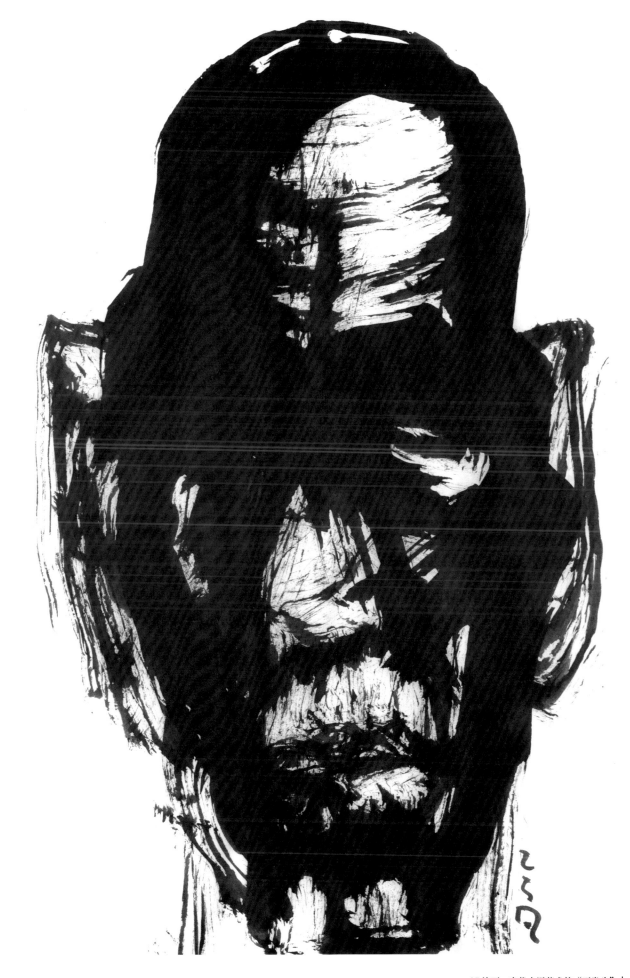

实验水墨
绘画材料：宣纸，水墨

Experimental Ink Wash Painting
Artwork: Rice paper, Ink
33cm×46cm
2023

实验水墨
绘画材料：速写纸，水墨

Experimental Ink Wash Painting
Artwork: Sketching paper, Ink
21cm × 28cm
2023

基于左图的 AI 衍生数字绘画
媒材模拟：速写纸，水墨

AI-derived digital painting based on the painting (Left)
Simulated Material: Sketching paper, Ink
4266px × 4266px
2023

实验水墨

绘画材料：素描纸，水墨

Experimental Ink Wash Painting
Artwork:Sketch paper, Ink
21cm×28cm
2023

基于左图的 AI 衍生数字绘画

媒材模拟：素描纸，水墨

AI-derived digital painting based on the painting (Left)
Simulated Material:Sketch paper, Ink
4266px×4266px
2023

基于左图的 AI 衍生数字绘画
媒材模拟：素描纸，水墨

AI-derived digital painting based on the painting (Left)
Simulated Material:Sketch paper, Ink
4266px × 4266px
2023

基于左图的 AI 衍生数字绘画
媒材模拟：素描纸，水墨

AI-derived digital painting based on the painting (Left)
Simulated Material:Sketch paper, Ink
4266px × 4266px
2023

实验水墨
绘画材料：牛皮纸，水墨

Experimental Ink Wash Painting
Artwork: Kraft paper, Ink
55cm×60cm
2023

基于左图的 AI 衍生数字绘画
媒材模拟：牛皮纸，水墨

AI-derived digital painting based on the painting (Left)
Simulated Material：Kraft paper, Ink
2613px × 3764px
2023

实验水墨
绘画材料：宣纸，水墨

Experimental Ink Wash Painting
Artwork：Rice paper, Ink
33cm × 46cm
2023

基于左图的 AI 衍生数字绘画
媒材模拟：宣纸，水墨

AI-derived digital painting based on the painting (Left)
Simulated Material: Rice paper, Ink
2480px × 3507px
2023

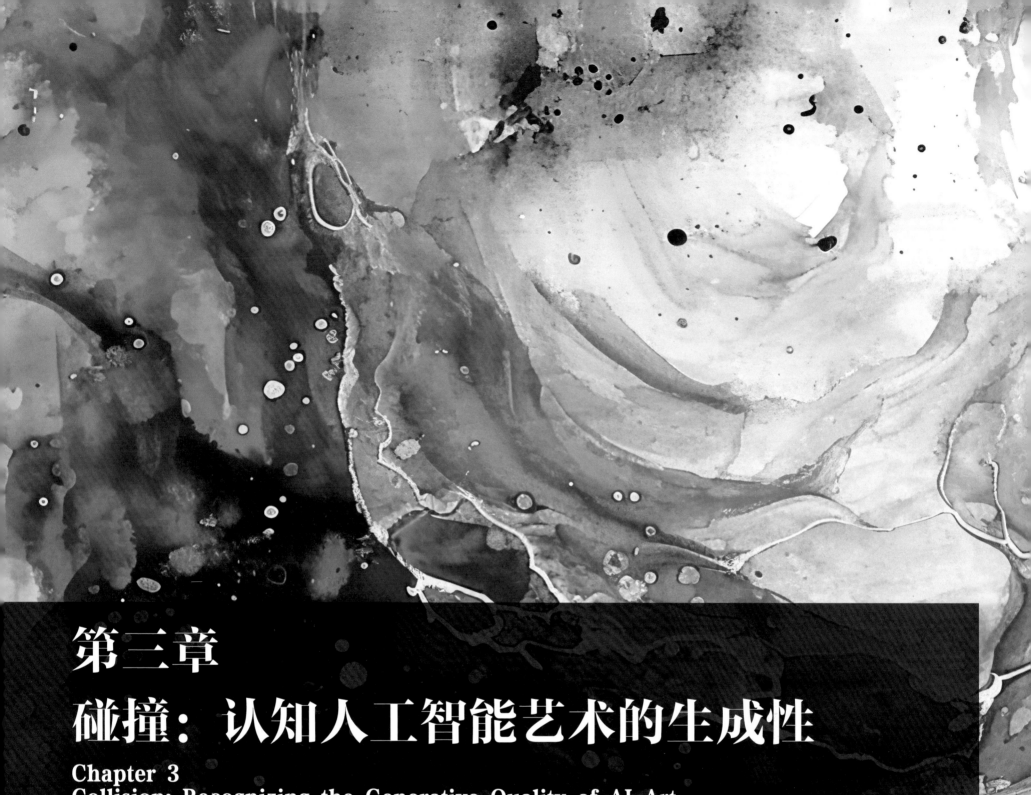

第三章
碰撞：认知人工智能艺术的生成性

Chapter 3
Collision: Recognizing the Generative Quality of AI Art

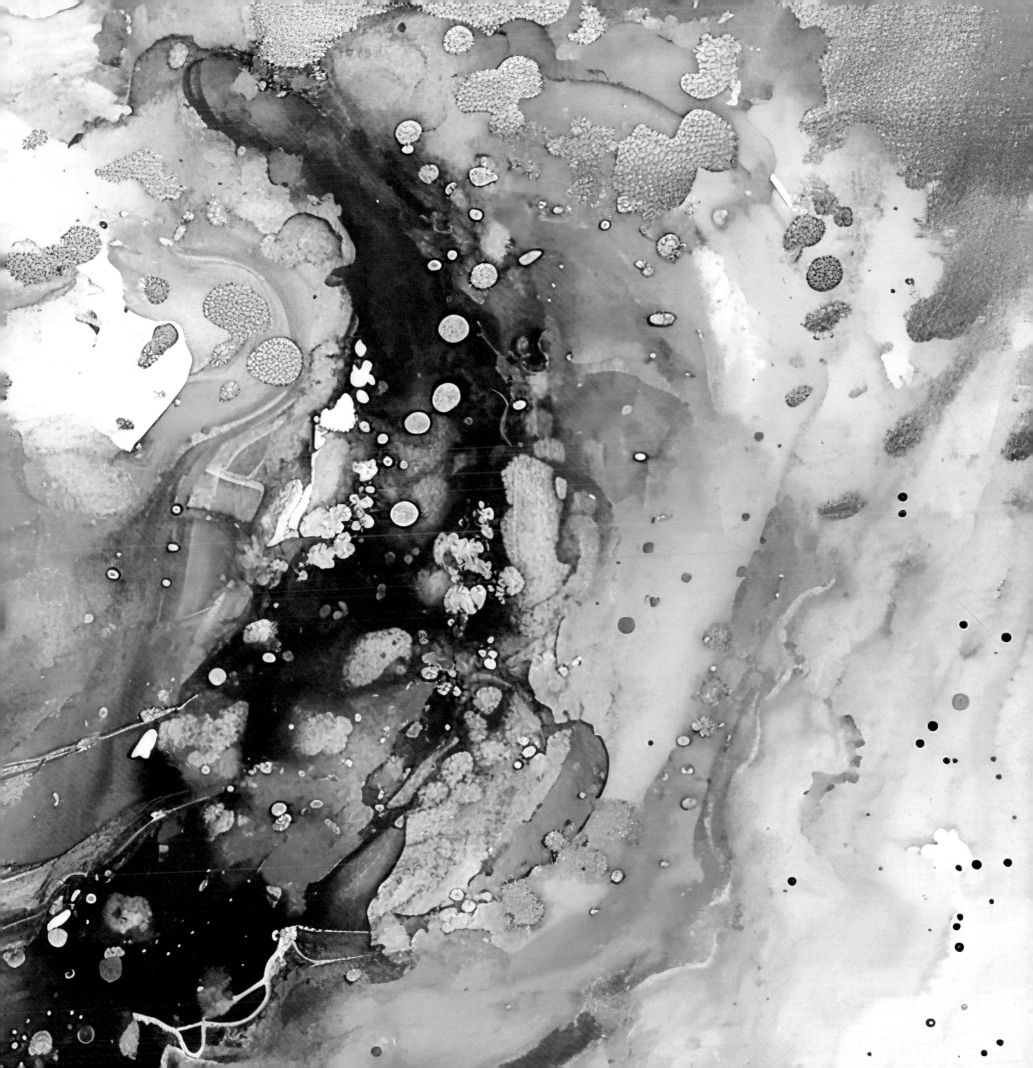

采访07：人工智能在艺术行业内不仅带来了生产效率的提升，也引发了以"艺术家必将被工具取代"为代表的焦虑与隐忧，对此您的看法是？

凡： 我是学传统工艺美术出身的，经历了从胶片时代到数字时代，从赛璐珞动画到"无纸化"创作形态等一系列转变。不得不承认，生成式人工智能绘画的出现及其以"月份"为单位的发展态势无疑令我感到惊奇，但随之而来的更多是惊喜。作为六零后，我非常有幸能够再次身临技术的转型与升级，并对人工智能绘画的未来发展持乐观态度。为什么乐观呢？因为它可以大大缩减艺术创作中重复性、机械式的工作环节所耗费的时间与人力成本，当这些环节由人工智能"代劳"后，艺术家能够将创作的重心真正转向创意开发的环节，正如工业化虽然造成大批工人失业，但实则推动人类文明向前迈进了一大步一般，我个人认为作为艺术创作者，我们与其惧怕人工智能，倒不如以开放的心态拥抱它，毕竟艺术最本质的内核还是在于创造"已知"之外的"未知"之境，传达诉诸心灵的精神力量。人工智能的出现可以帮助人类在短时间内完成大量的信息搜集、筛选与梳理工作，为我们即兴的创意提供多元形态的预演方式，让创意得以高效率地实现诉诸感官的直观呈现。相应的，省下来的时间则可以用于人类艺术家进行更具原创性与突破性的创意开发，深入探索贴合时代图景，映射人类共同理想的母题。从这个意义上说，人工智能的广泛应用不仅有益于艺术创作在思想、情感深度上的提升，更有益于引导艺术家将判断作品质量的标准更多地放在提升"原创力"上。因此就我个人的观点而言，我期待看到未来人类艺术家在人工智能技术支持下，突破传统艺术表达范式，让艺术创作在形式与内容上多元共生，百花齐放。

Chapter 3 Collision: Recognizing the Generative Quality of AI Art

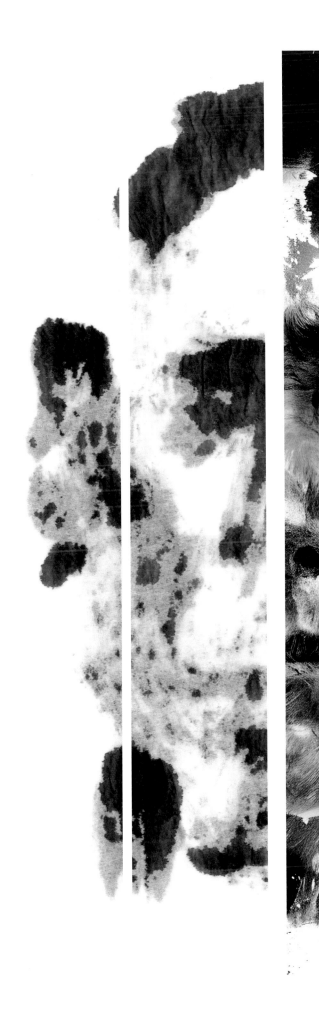

基于左图的 AI 衍生数字绘画
媒材模拟：速写纸，水墨

AI-derived digital painting based on the painting (Left)
Simulated Material: Sketching paper, Ink
928px × 1232px
2023

实验水墨
绘画材料：餐巾纸，水墨

Experimental Ink Wash Painting
Artwork: Napkin paper, Ink
21cm × 28cm
2023

Q7: Artificial intelligence has not only improved production efficiency within the art industry but has also sparked anxieties and worries represented by the notion that "artists will inevitably be replaced by tools." What is your opinion on this matter?

Fan: My artistic training initially started from traditional arts and crafts learning, throughout the years, I had experienced a series of transitions from the era of film to the digital age, from traditional animation to "paperless" creative forms, and so forth. I must admit that the emergence of generative AI art and its development trends on a monthly frequency have indeed astonished me, but more importantly, instead of a sense of crisis, they have brought me a sense of delight. As someone born in the 1960s, I am fortunate to once again witness the transformation and upgrade of technology, and I hold an optimistic attitude towards the future development of AI art. Why am I optimistic? Because it can significantly reduce the time and human resources costs spent on repetitive and mechanical tasks in artistic production process. When these tasks are taken over by artificial intelligence, artists can get themselves fully focused on the creative development process. Just as industrialization has led to mass unemployment but ultimately propelled human civilization forward, I personally believe that as artists, rather than fearing artificial intelligence, we should embrace it with an open mindset. After all, the essence of art lies in exploring the unknown beyond the known and conveying the spiritual power that resonates with the soul.The emergence of artificial intelligence can help humans complete a large amount of information collection, filtering, and organization in a short period of time, offering diverse forms of visualization for our impromptu creative ideas, allowing them to be efficiently realized in an intuitive presentation that appeals to the senses. Accordingly, the time saved can be used for human artists to develop more original and groundbreaking creative work, to deeply explore themes that mirror the era and

Chapter 3 Collision: Recognizing the Generative Quality of AI Art

reflect common human ideals. In this sense, the widespread application of artificial intelligence is not only beneficial for enhancing the depth of thought and emotion in artistic creation but also for guiding artists to place more emphasis on improving "originality" as the standard for judging the quality of work. Therefore, from my personal perspective, I look forward to seeing human artists break through traditional artistic expression paradigms with the support of artificial intelligence technology, allowing art creation to flourish in form and content, and to blossom diversely.

实验水墨
绘画材料：素描纸，水墨

Experimental Ink Wash Painting
Artwork:Sketch paper, Ink
21cm × 28cm
2023

基于上图的 AI 衍生数字绘画
媒材模拟：素描纸，水墨

AI-derived digital painting based on the painting (Upper)
Simulated Material:Sketch paper, Ink
928px × 1232px
2023

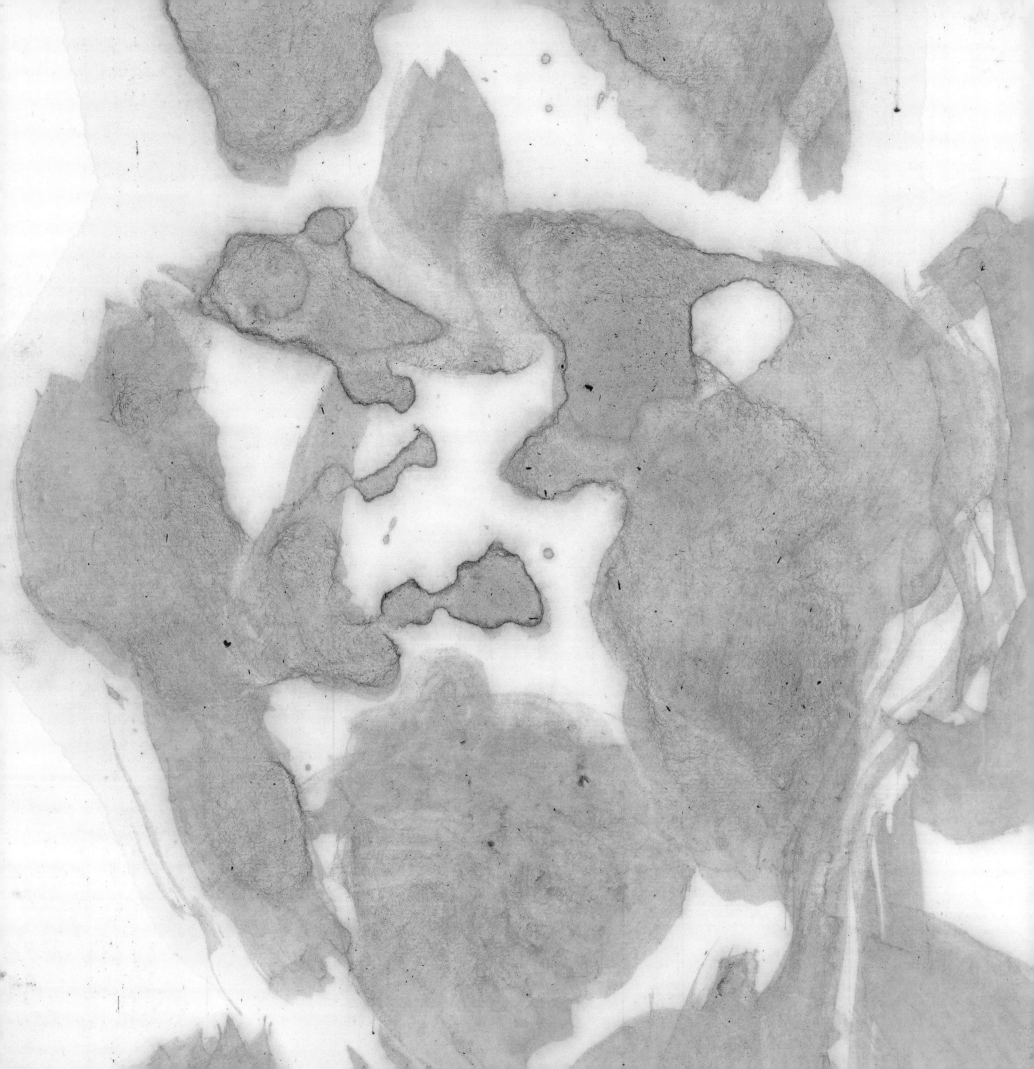

采访 08：您如何理解人工智能"生成"一幅画与人类"创作"一幅画在底层逻辑上的异同？

凡： 总体而言，人工智能"生成式"的绘画主要依赖算法和大数据模型驱动，其生成机制遵循预设定义的规则和算法中编码的模式，旨在通过反复训练实现某种视觉风格与审美范式的精确复制。人类"创作"一幅画诚然也是建立在日复一日苦练基本功的基础上，但决定一幅画作质量高低的往往不仅在于其在"机械复制"层面表现出的功底，而更在于在画作中呈现创作者的直觉、情感和个人经验在"某个当下"与特定环境所碰撞出的"临场感"。此外，当我们回归绘画乃至创作的目的，人工智能的目的仅仅在于有效率地生成合乎标准的作品，但对人类而言并非如此，创作的过程有时比结果更为重要，因此我主张人类艺术家切忌被人工智能的节奏"卷着走"，只注重结果而忽视过程中原创的重要性。这个观点在我自己创作的动画短片，如《秋实》《立秋》，尤其是《最后的舞台》中都有探讨。我想对于创作者，如果肉体走得过快，而灵魂没有跟上，即作品只有夺目的奇观而缺乏精神的深度，无法与观者的心灵产生共振，这实则是一件很悲哀的事，它将让我们的创作偏离其原初意义的旨归。这也是为什么我在日常教学与个人创作中，更重视微妙的、主观的、感性的表达，以人类直觉、创造力和对个体表达的追求作为创作的主要驱动力的原因之一。

实验水墨
绘画材料：素描纸，水墨

Experimental Ink Wash Painting
Artwork: Sketch paper, Ink
21cm × 28cm
2023

第三章　碰撞：认知人工智能艺术的生成性

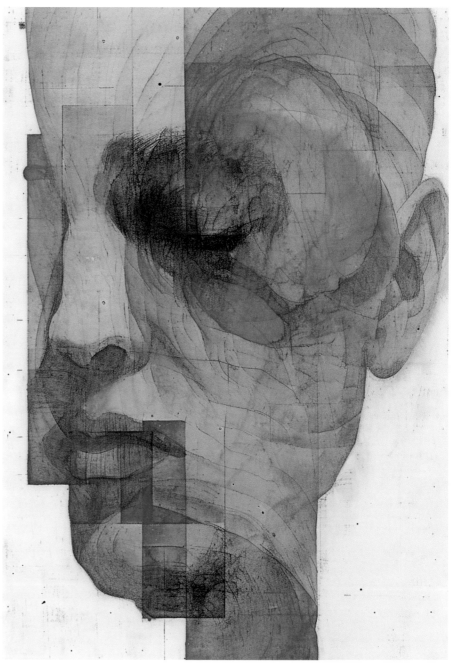

基于 178 页作品的 AI 衍生数字绘画
媒材模拟：素描纸，水墨

AI-derived digital painting based on the painting (P178)
Simulated Material:Sketch paper, Ink
928px × 1232px
2023

基于 178 页作品的 AI 衍生数字绘画
媒材模拟：素描纸，水墨

AI-derived digital painting based on the painting (P178)
Simulated Material:Sketch paper, Ink
928px × 1232px
2023

Q8: What do you think is the significant underlying logic differences and similarities between AI "generating" a painting and humans "creating" a painting?

Fan: Overall, AI "generative" painting mainly relies on algorithms and large-scale data models. Its generation mechanism follows preset rules and patterns encoded in algorithms, aiming to achieve precise replication of certain visual styles and aesthetic paradigms through repeated training. While human "creation" of a painting is indeed based on the foundation of practicing basic skills day after day, the quality of a painting is often determined not only by its proficiency in "mechanical reproduction" but also by the collision of the creator's intuition, emotions, and personal experiences with a specific environment in a particular moment. In addition, when we return to the purpose of painting and even creation itself, the purpose of artificial intelligence is merely to efficiently generate works that meet certain standards. However, for humans, it is not the same. Sometimes, the process of creation is more important than the result. Therefore, I advocate that human artists should not be swept away by the sudden rushing-in of artificial intelligence, focusing only on the result and neglecting the importance of originality in the process. This viewpoint is discussed in many animated short films I directed during previous years, such as *Qiu Shi* and *Li Qiu* especially in *The Last Stage*. For creators, if the body moves too fast while the soul fails to keep up, their works are very likely to have dazzling spectacles but lack spiritual depth and fail to resonate with the viewer's mind and soul, which would be quite tragic. It would lead our creation astray from its original purpose. That's why in my daily teaching and personal creation, I emphasize subtle, subjective, and emotional expression more, I encourage my students to infuse their intuition, creativity, and pursuit of individual expression into their works and see them as the main driving forces of creation.

第三章 碰撞：认知人工智能艺术的生成性

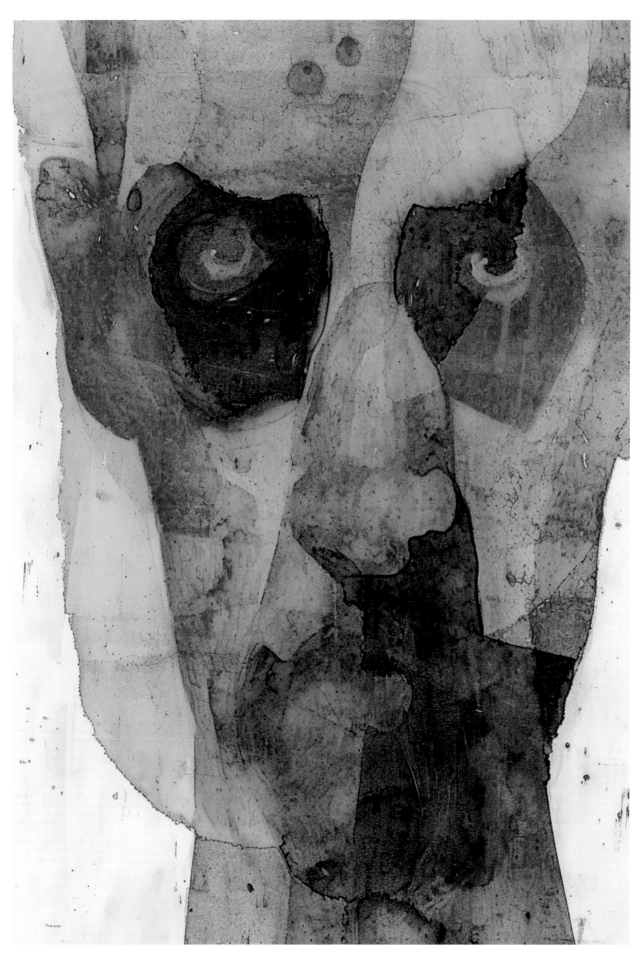

基于 178 页作品的 AI 衍生数字绘画
媒材模拟：素描纸，水墨

AI-derived digital painting based on the painting (P178)
Simulated Material: Sketch paper, Ink
928px × 1232px
2023

采访09：您提到"主张与人工智能成为创作上的伙伴"，您认为未来人类在艺术创作领域应当如何与人工智能实现通力合作？

凡： 实现通力合作，某种意义上就是整合人类与人工智能在艺术创作领域的优势，让二者的优势"做加法"。比如在创意开发阶段，艺术家可以输入参数或初始概念，由人工智能算法生成多元的、直观的内容文本，以辅助进一步的创意开发；在媒介融合上，人类艺术家可以尝试在人工智能技术支持下使用虚拟现实（VR）、增强现实（AR）或混合现实（MR）等新兴技术以丰富作品的表现形态，如人机互动装置的开发等；本书中所尝试的某种意义上可被视为"个性化风格学习"，即以某一艺术家或某一创作流派的作品为数据库，训练人工智能模型学习特定的构图、透视、造型、用色等方面的章法与偏好，使其能够输出同数据库在风格与审美上高度一致的作品。同时艺术家在创作过程中应用人工智能图像生成能力与模式识别功能的同时，还应当注重人工智能大数据分析的优势，在人机交互的动态过程中反思自身的创作模式与审美范式是否有可以实现自我突破的空间，促使艺术创作突破个体创造力的局限，让艺术创作的"过程"和"目标"都获得更具实验性与开放性的阐释空间，在创作与阐释的过程中充分探索人类艺术家直觉与情感的表达方式与表现手法，为艺术创作注入无限的可能性与多样性。当然，实现通力合作还需形成一套新的伦理规范，包括对AI创作内容的版权归属、艺术家与AI合作关系的界定等。

第三章　碰撞：认知人工智能艺术的生成性

基于 178 页作品的 AI 衍生数字绘画
媒材模拟：素描纸，水墨

AI-derived digital painting based on the painting (P178)
Simulated Material:Sketch paper, Ink
928px × 1232px
2023

基于 178 页作品的 AI 衍生数字绘画
媒材模拟：素描纸，水墨

AI-derived digital painting based on the painting (P178)
Simulated Material:Sketch paper, Ink
928px × 1232px
2023

Q9: You mentioned "advocating for collaboration with artificial intelligence". How do you think humans should cooperate with artificial intelligence in the field of artistic creation in the future?

Fan: Achieving collaboration essentially means integrating the strengths of humans and artificial intelligence in the field of artistic creation, allowing the advantages of both to complement each other. For example, in the stage of brain storming an idea, artists can input parameters or initial concepts, and artificial intelligence algorithms can generate diverse and intuitive content to assist further creative development. In terms of media integration, human artists can explore the use of emerging technologies such as virtual reality (VR), augmented reality (AR), or mixed reality (MR) with the support of artificial intelligence technology to enrich the presentation forms of works, such as the development of human-machine interactive installations. What we attempted in this book can be seen as a form of "personalized style learning", that is the works of a particular artist or a specific creative genre serve as a database to train artificial intelligence models to learn specific conventions and preferences in composition, perspective, form, and color, enabling them to output works highly consistent with the style and aesthetics of the database. While artists apply artificial intelligence's image generation capabilities and pattern recognition functions in the creative process, they should also focus on the advantages of AI's big data analysis. In the dynamic process of human-machine interaction, they should reflect on whether their creative models and aesthetic paradigms have room for self-breakthrough, pushing artistic creation beyond the limits of individual creativity. This approach allows not only the "process" but also the "goal" of artistic creation to gain more experimental and theoretical value. During the creation and interpretation process, it fully explores the ways and methods in which human artists express their intuition and emotion, injecting endless possibilities and diversity into artistic creation. Of course, achieving close cooperation also requires the formation of a new set of ethical norms, including copyright ownership of AI-created content, the definition of the relationship between artists and AI, etc.

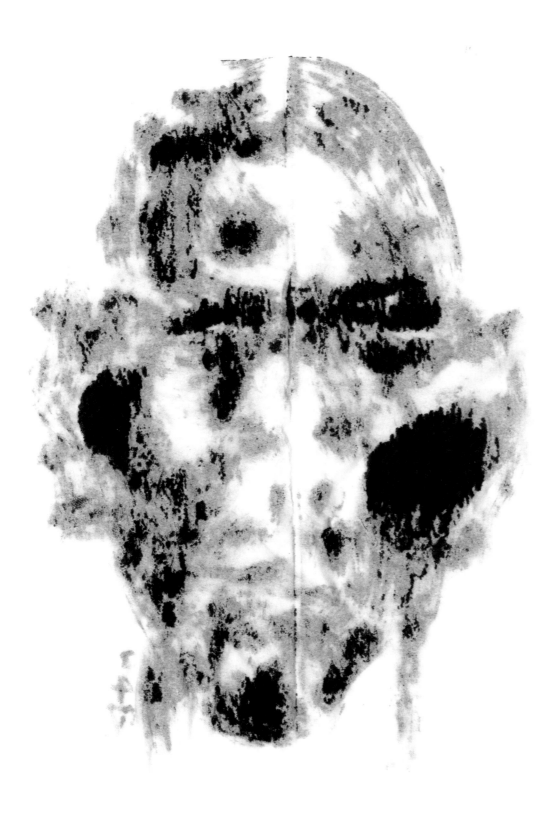

实验水墨
绘画材料：餐巾纸，水墨

Experimental Ink Wash Painting
Artwork:Napkin paper, Ink
10cm×16cm
2023

基于左图的 AI 衍生数字绘画
媒材模拟：速写纸，水墨

AI-derived digital painting based on the painting (Left)
Simulated Material: Sketching paper, Ink
928px × 1232px
2023

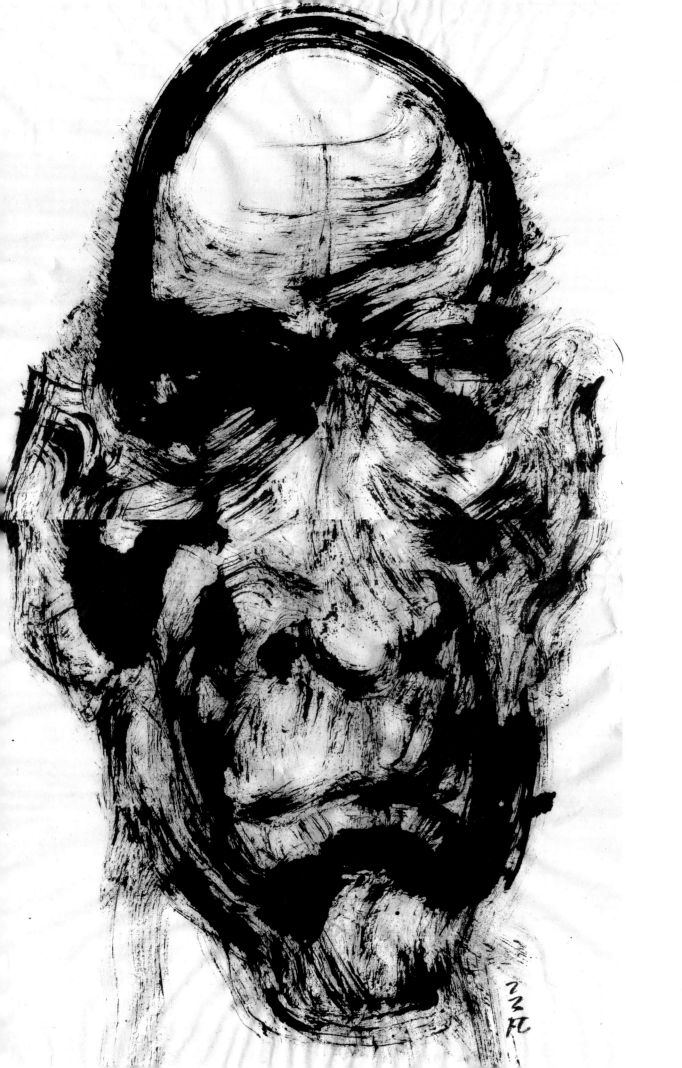

实验水墨
绘画材料：宣纸，水墨

Experimental Ink Wash Painting
Artwork: Rice paper, Ink
55cm×60cm
2023

基于左图的 AI 衍生数字绘画
媒材模拟：宣纸，水墨

AI-derived digital painting based on the painting (Left)
Simulated Material: Rice paper, Ink
928px × 1232px
2023

基于左图的 AI 衍生数字绘画
媒材模拟：宣纸，水墨

AI-derived digital painting based on the painting (Left)
Simulated Material: Rice paper, Ink
928px × 1232px
2023

基于 188 页图的 AI 衍生数字绘画
媒材模拟：宣纸，水墨

AI-derived digital painting based on the painting (P188)
Simulated Material: Rice paper, Ink
928px × 1232px
2023

基于 188 页图的 AI 衍生数字绘画
媒材模拟：宣纸，水墨

AI-derived digital painting based on the painting (P188)
Simulated Material: Rice paper, Ink
928px × 1232px
2023

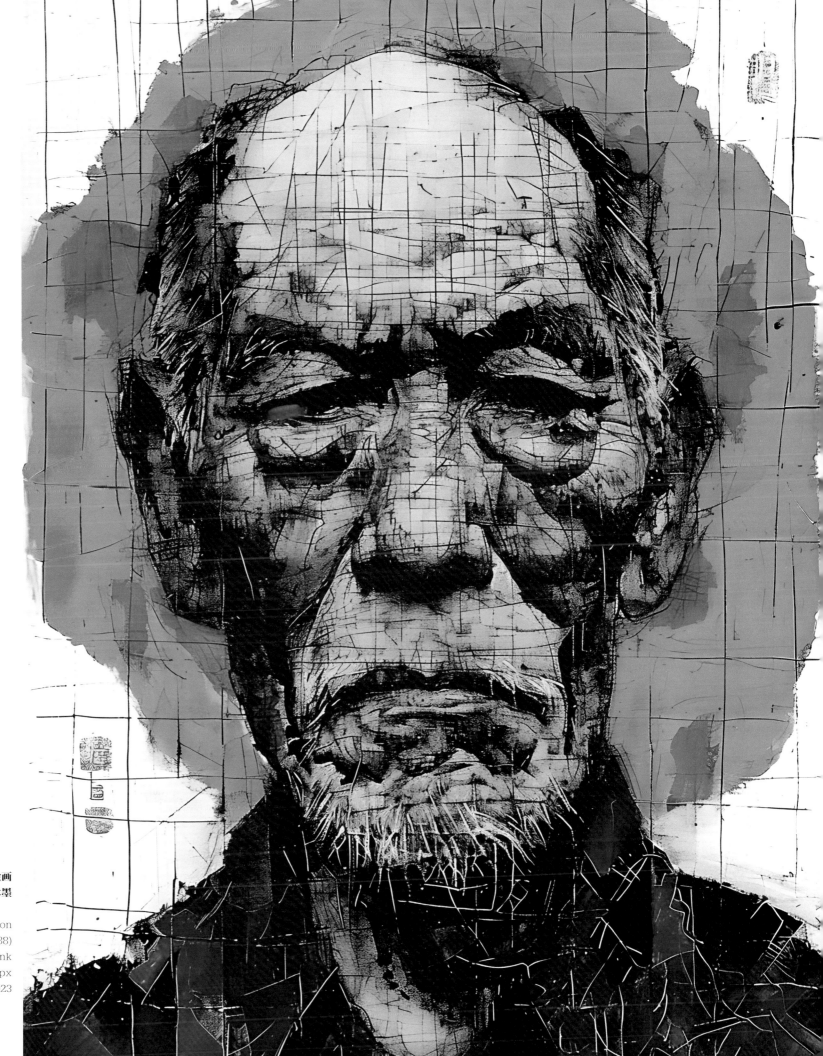

基于 188 页图的 AI 衍生数字绘画
媒材模拟：宣纸，水墨

AI-derived digital painting based on
the painting (P188)
Simulated Material: Rice paper, Ink
928px × 1232px
2023

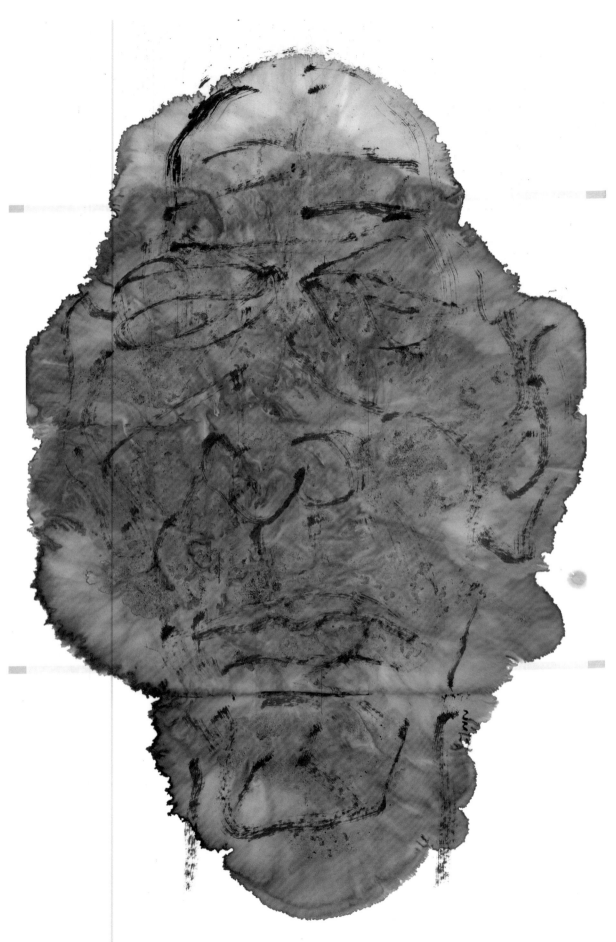

实验水墨
绘画材料：素描纸，水墨

Experimental Ink Wash Painting
Artwork: Sketch paper, Ink
21cm × 28cm
2023

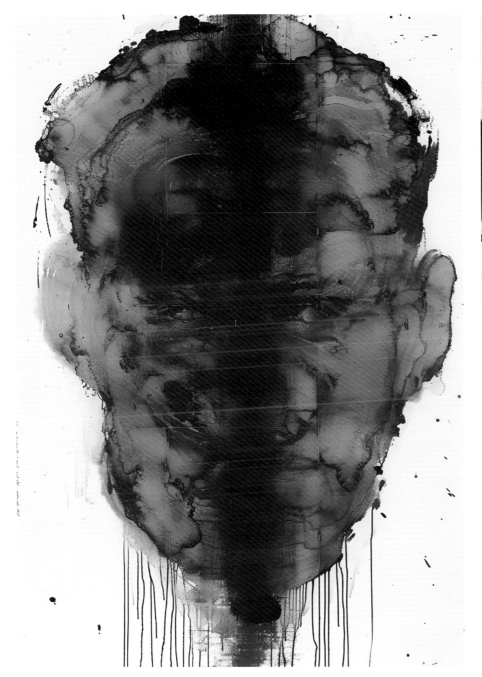

基于左图的 AI 衍生数字绘画
媒材模拟：素描纸，水墨

AI-derived digital painting based on the painting (Left)
Simulated Material: Sketch paper, Ink
928px × 1232px
2023

基于左图的 AI 衍生数字绘画
媒材模拟：素描纸，水墨

AI-derived digital painting based on the painting (Left)
Simulated Material: Sketch paper, Ink
928px × 1232px
2023

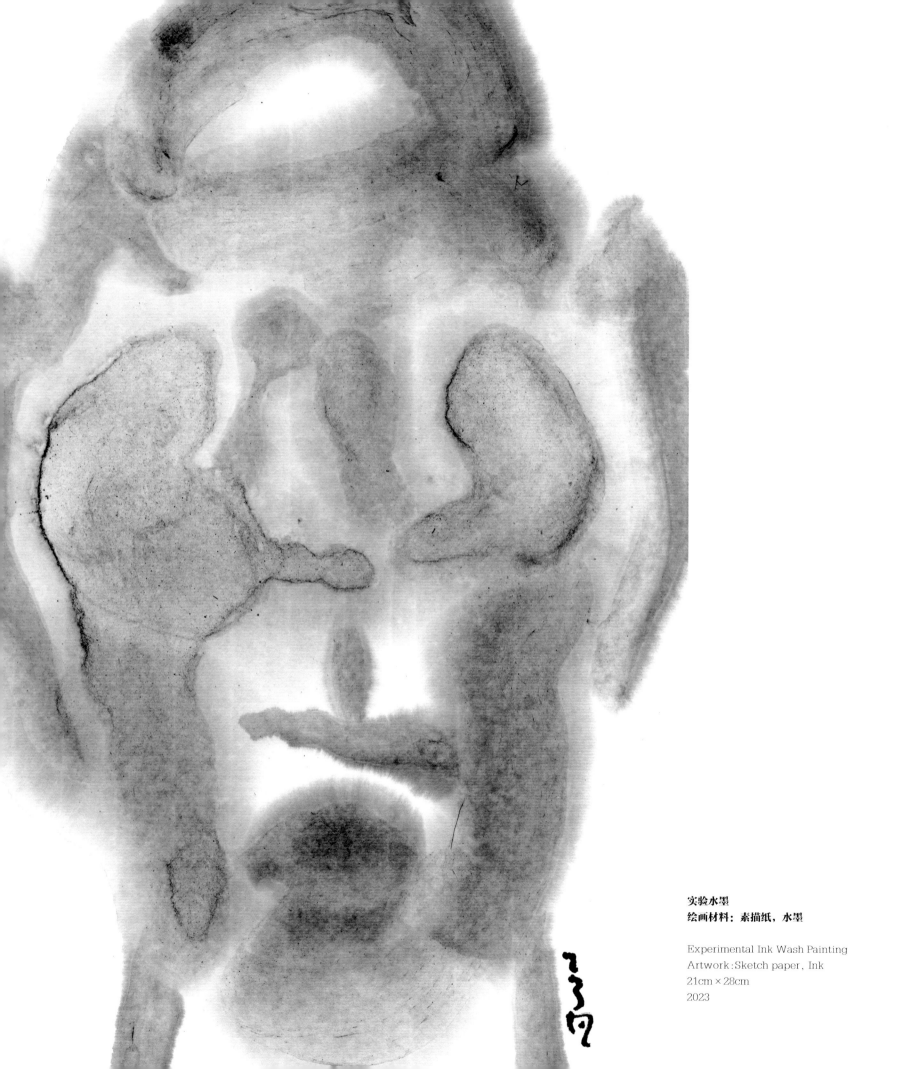

实验水墨
绘画材料：素描纸，水墨

Experimental Ink Wash Painting
Artwork:Sketch paper, Ink
21cm×28cm
2023

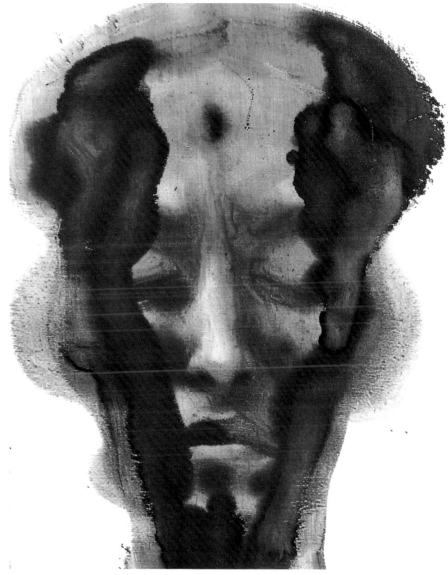

基于左图的 AI 衍生数字绘画
媒材模拟：素描纸，水墨

AI-derived digital painting based on the painting (Left)
Simulated Material: Sketch paper, Ink
928px × 1232px
2023

基于左图的 AI 衍生数字绘画
媒材模拟：素描纸，水墨

AI-derived digital painting based on the painting (Left)
Simulated Material: Sketch paper, Ink
928px × 1232px
2023

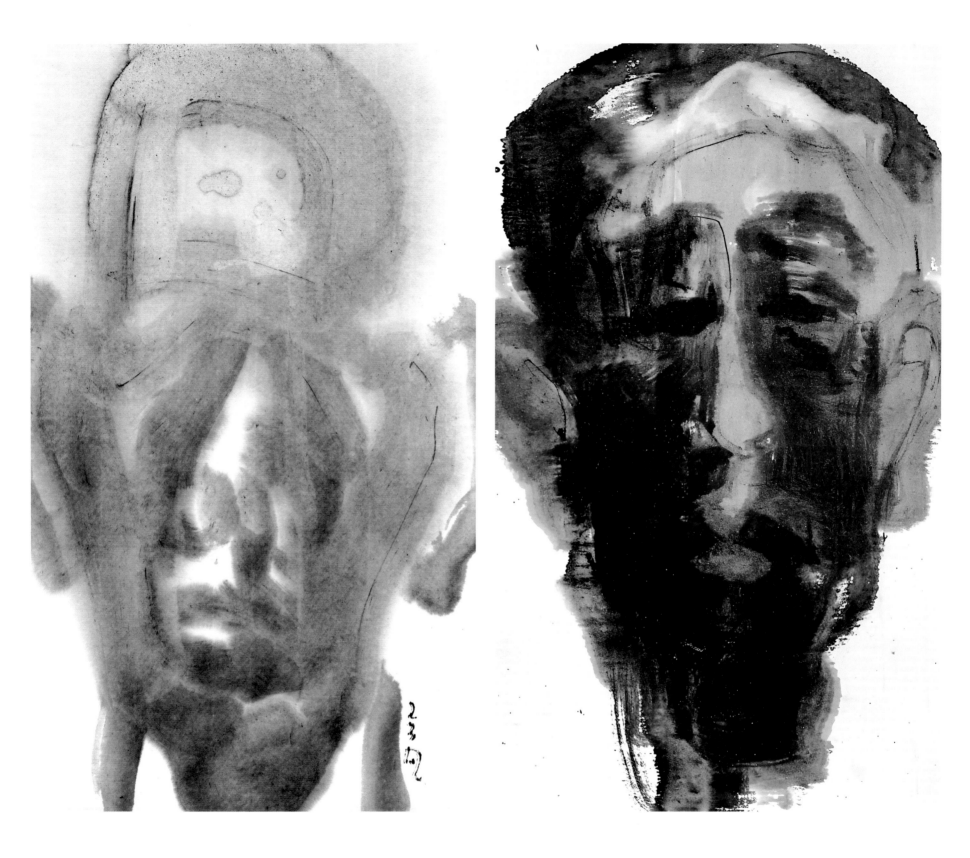

实验水墨
绘画材料：素描纸，水墨

Experimental Ink Wash Painting
Artwork：Sketch paper, Ink
21cm×28cm
2023

基于左图的 AI 衍生数字绘画
媒材模拟：素描纸，水墨

AI-derived digital painting based on the painting (Left)
Simulated Material：Sketch paper, Ink
928px×1232px
2023

基于左图的 AI 衍生数字绘画
媒材模拟：素描纸，水墨

AI-derived digital painting based on the painting (Left)
Simulated Material:Sketch paper, Ink
928px × 1232px
2023

基于左图的 AI 衍生数字绘画
媒材模拟：素描纸，水墨

AI-derived digital painting based on the painting (Left)
Simulated Material:Sketch paper, Ink
928px × 1232px
2023

AI 绘画：当代水墨艺术的"正发生" | 197

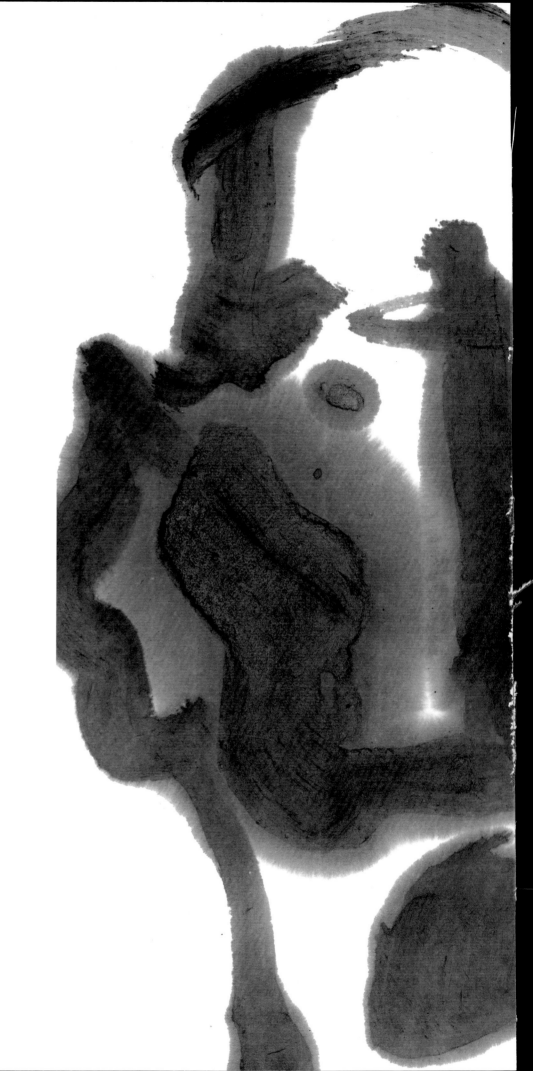

实验水墨
绘画材料：素描纸，水墨

Experimental Ink Wash Painting
Artwork:Sketch paper, Ink
21cm×28cm
2023

基于左图的 AI 衍生数字绘画
媒材模拟：素描纸，水墨

AI-derived digital painting based on the painting (Left)
Simulated Material: Sketch paper, Ink
928px × 1232px
2023

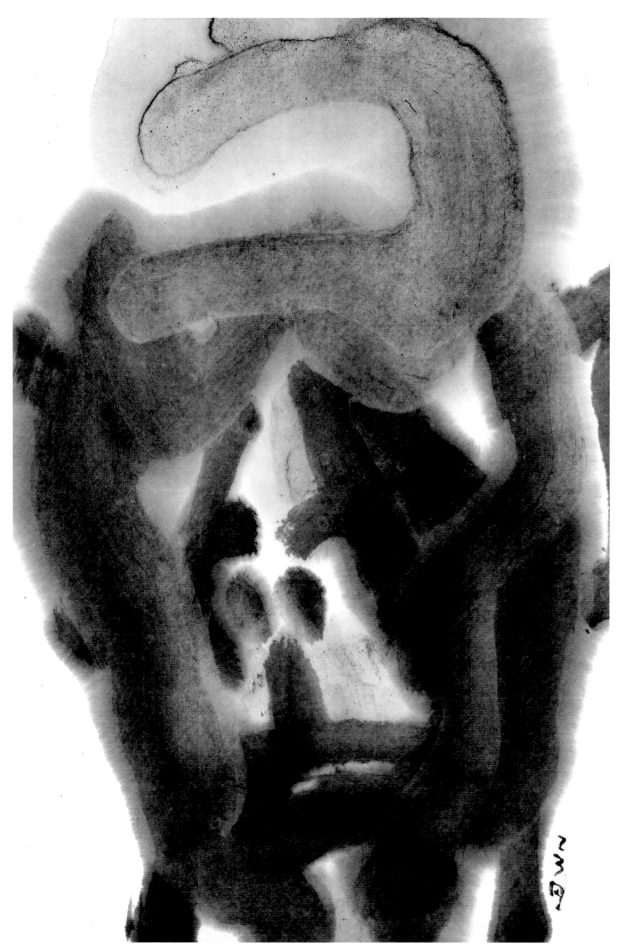

实验水墨
绘画材料：素描纸，水墨

Experimental Ink Wash Painting
Artwork:Sketch paper, Ink
21cm×28cm
2023

基于左图的 AI 衍生数字绘画
媒材模拟：素描纸，水墨

AI-derived digital painting based on the painting (Left)
Simulated Material: Sketch paper, Ink
1024px × 1024px
2023

"作为艺术创作者，

我们与其惧怕人工智能，

倒不如以开放的心态拥抱它，

毕竟艺术最本质的内核还是在于创造'已知'之外的'未知'之境，

传达诉诸心灵的精神力量。"

"I personally believe that as artists, rather than fearing artificial intelligence,

we should embrace it with an open mindset.

After all, the essence of art lies in exploring the unknown beyond the known

and conveying the spiritual power that resonates with the soul."

实验水墨
绘画材料：素描纸，水墨

Experimental Ink Wash Painting
Artwork: Sketch paper, Ink
21cm × 28cm
2023

基于左图的 AI 衍生数字绘画
媒材模拟：素描纸，水墨

AI-derived digital painting based on the painting (Left)
Simulated Material: Sketch paper, Ink
1024px × 1024px
2023

实验水墨
绘画材料：素描纸，水墨

Experimental Ink Wash Painting
Artwork:Sketch paper, Ink
21cm×28cm
2023

基于左图的 AI 衍生数字绘画
媒材模拟：素描纸，水墨

AI-derived digital painting based on the painting (Left)
Simulated Material:Sketch paper, Ink
1024px×1024px
2023

实验水墨
绘画材料：素描纸，水墨
Experimental Ink Wash Painting
Artwork：Sketch paper, Ink
21cm×28cm
2023

基于左图的 AI 衍生数字绘画
媒材模拟：素描纸，水墨
AI-derived digital painting based on the painting (Left)
Simulated Material：Sketch paper, Ink
4266px×4266px
2023

**基于左图的 AI 衍生数字绘画
媒材模拟:素描纸，水墨**

AI-derived digital painting
based on the painting (Left)
Simulated Material:Sketch
paper, Ink
4266px × 4266px
2023

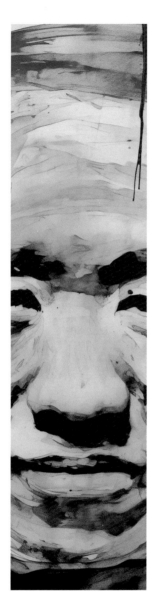

基于右图的 AI 衍生数字绘画
媒材模拟：牛皮纸，水墨

AI-derived digital painting based on the painting (Right)
Simulated Material: Kraft paper, Ink
1364px × 1665px
2023

实验水墨
绘画材料：牛皮纸，水墨

Experimental Ink Wash Painting
Artwork: Kraft paper, Ink
33cm × 46cm
2023

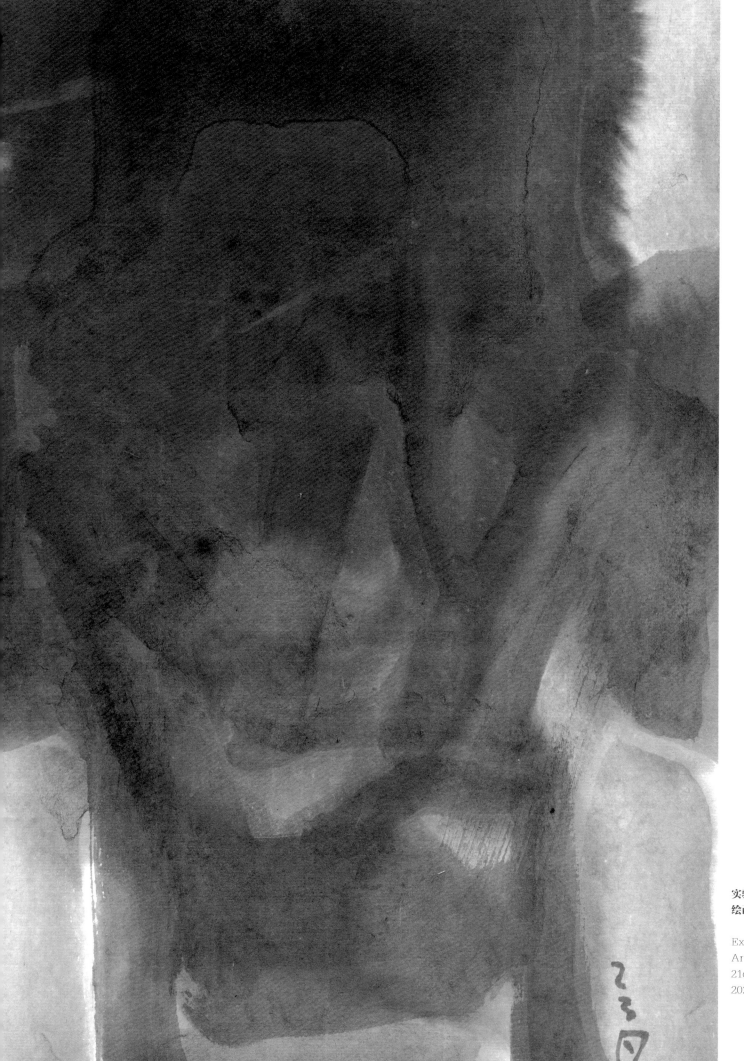

实验水墨
绘画材料：素描纸，水墨

Experimental Ink Wash Painting
Artwork:Sketch paper, Ink
21cm×28cm
2023

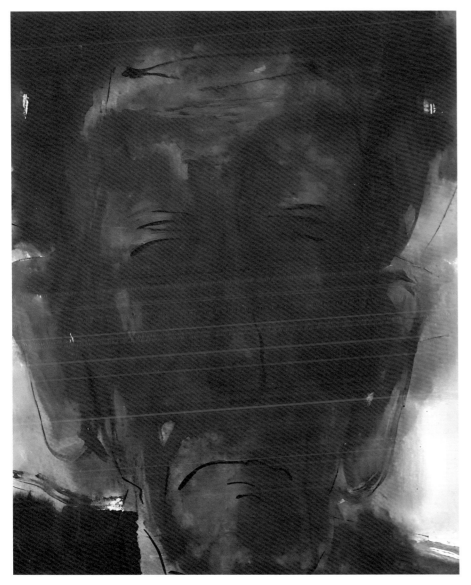

基于左图的 AI 衍生数字绘画
媒材模拟：素描纸，水墨

AI-derived digital painting based on the painting (Left)
Simulated Material:Sketch paper, Ink
1024px × 1024px
2023

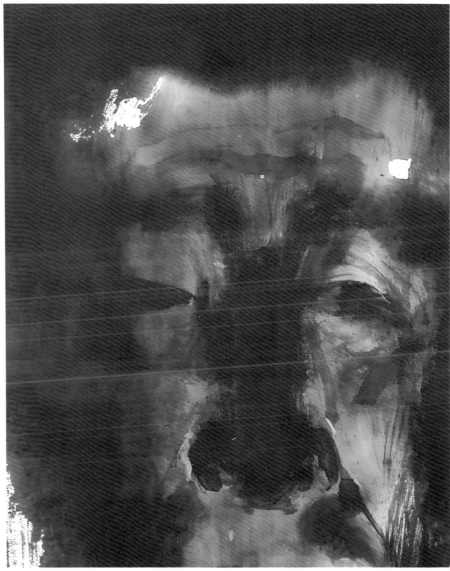

基于左图的 AI 衍生数字绘画
媒材模拟：素描纸，水墨

AI-derived digital painting based on the painting (Left)
Simulated Material:Sketch paper, Ink
1024px × 1024px
2023

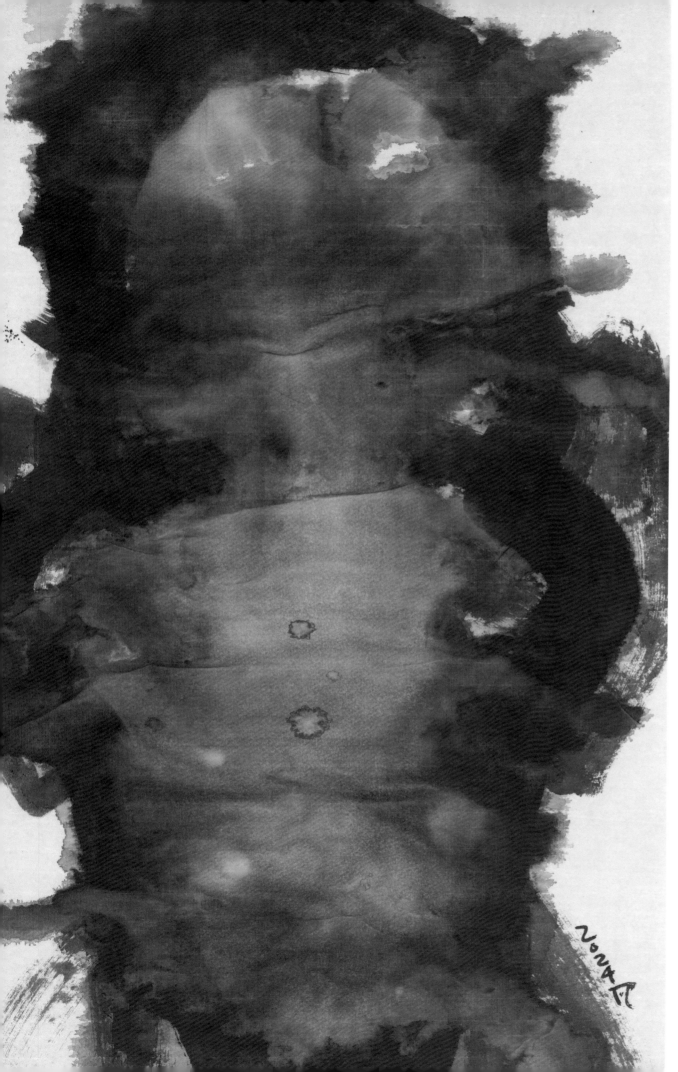

实验水墨
绘画材料：牛皮纸，水墨

Experimental Ink Wash Painting
Artwork：Kraft paper, Ink
55cm×60cm
2023

基于左图的 AI 衍生数字绘画
媒材模拟：牛皮纸，水墨

AI-derived digital painting based on the painting (Left)
Simulated Material: Kraft paper, Ink
928px × 1232px
2023

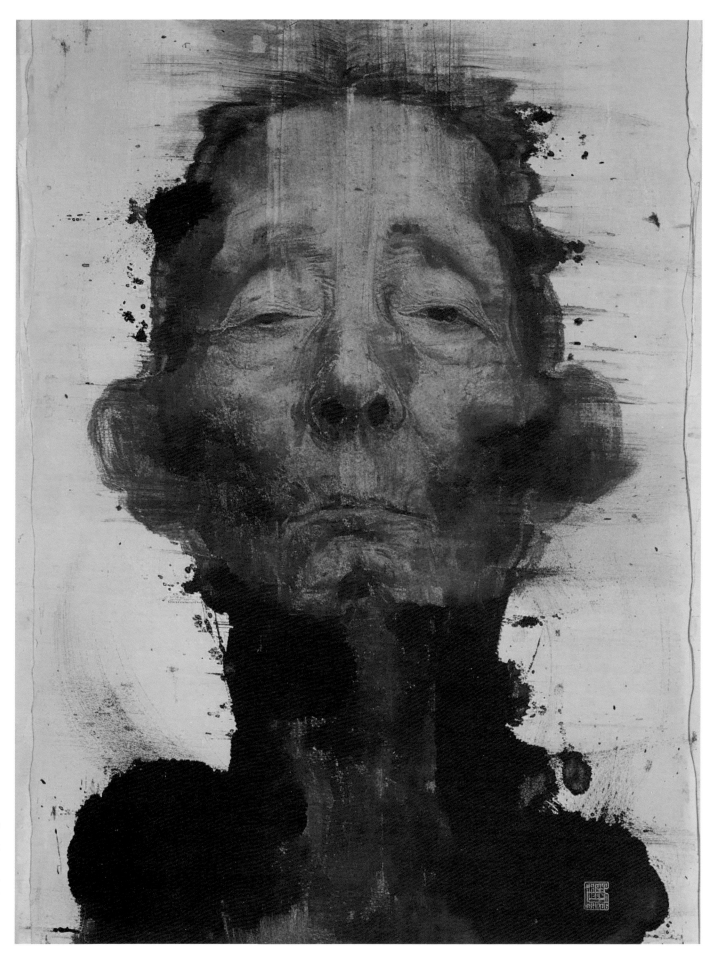

基于 212 页作品的 AI 衍生数字绘画
媒材模拟：牛皮纸，水墨

AI-derived digital painting based
on the painting (P212)
Simulated Material: Kraft paper, Ink
928px × 1232px
2023

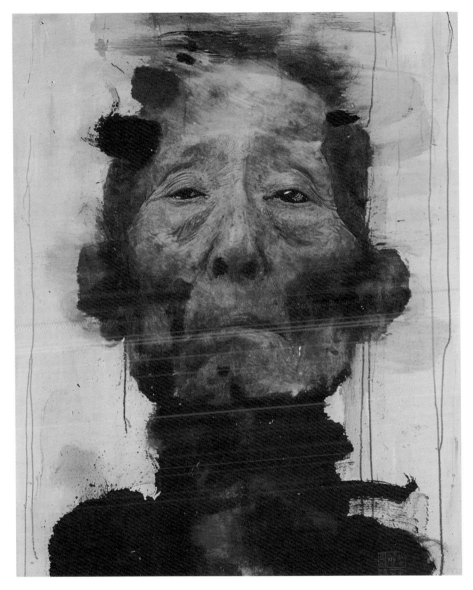

基于 212 页作品的 AI 衍生数字绘画
媒材模拟：牛皮纸，水墨

AI-derived digital painting based on the painting (P212)
Simulated Material:Kraft paper, Ink
928px × 1232px
2023

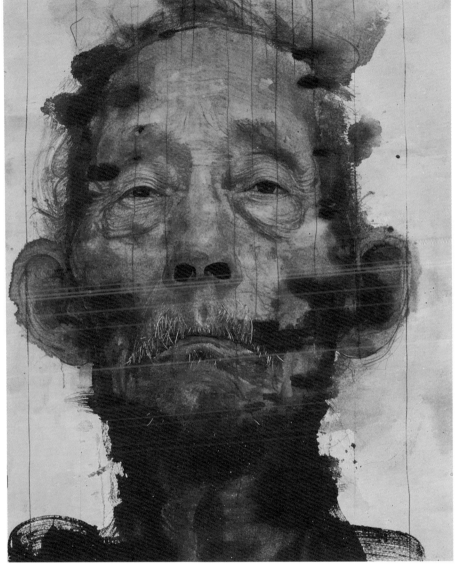

基于 212 页作品的 AI 衍生数字绘画
媒材模拟：牛皮纸，水墨

AI-derived digital painting based on the painting (P212)
Simulated Material:Kraft paper, Ink
928px × 1232px
2023

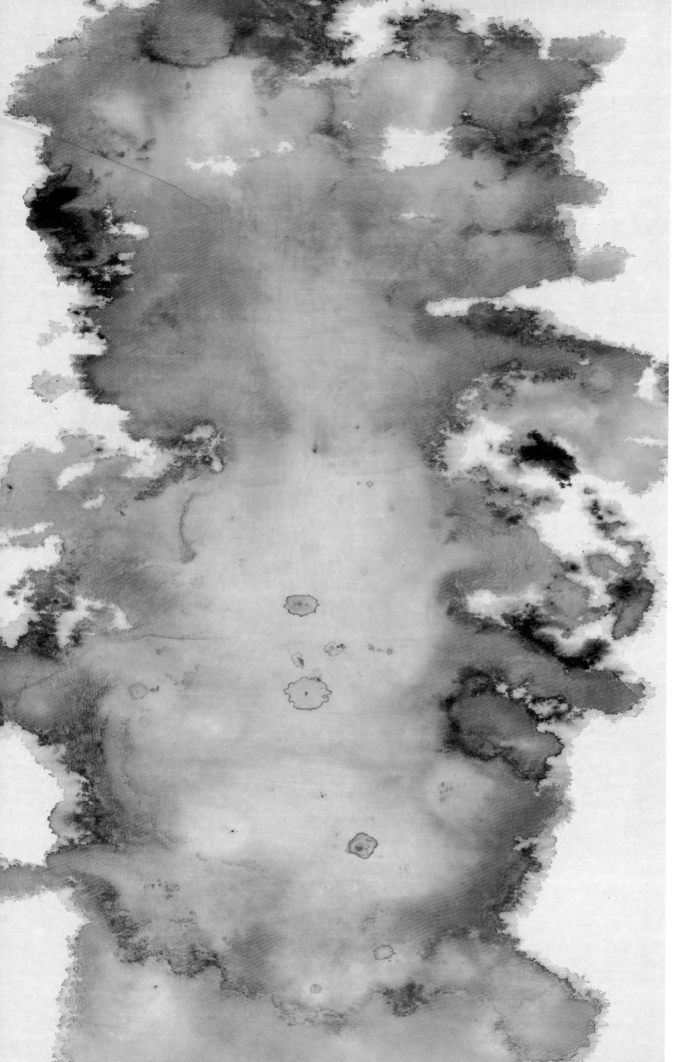

实验水墨
绘画材料：牛皮纸，水墨

Experimental Ink Wash Painting
Artwork:Kraft paper, Ink
55cm×60cm
2023

基于左图的 AI 衍生数字绘画
媒材模拟：牛皮纸，水墨

AI-derived digital painting based on the painting (Left)
Simulated Material: Kraft paper, Ink
928px × 1232px
2023

 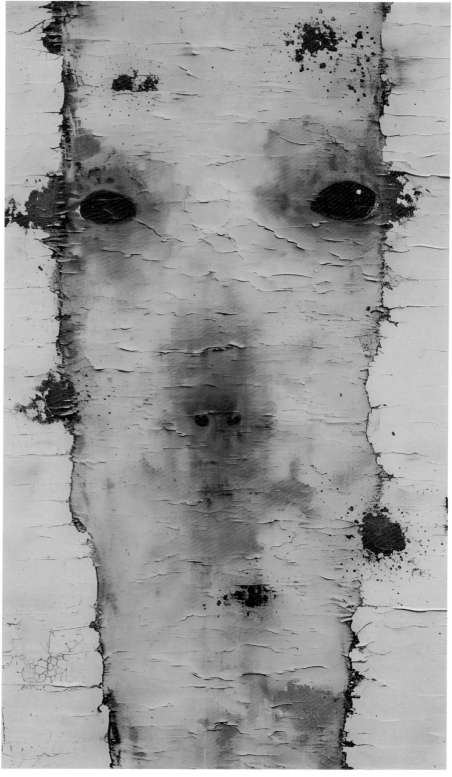

基于 216 页作品的 AI 衍生数字绘画
媒材模拟：牛皮纸，水墨

AI-derived digital painting based on the painting (P216)
Simulated Material:Kraft paper, Ink
928px × 1232px
2023

基于 216 页作品的 AI 衍生数字绘画
媒材模拟：牛皮纸，水墨

AI-derived digital painting based on the painting (P216)
Simulated Material:Kraft paper, Ink
928px × 1232px
2023

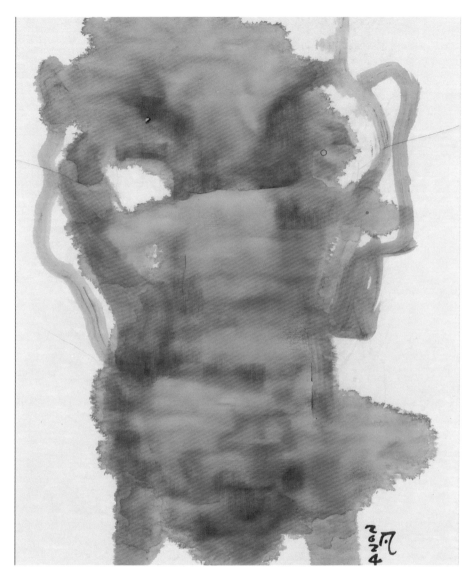

实验水墨
绘画材料：牛皮纸，水墨
Experimental Ink Wash Painting
Artwork: Kraft paper, Ink
55cm×60cm
2024

基于左图的 AI 衍生数字绘画
媒材模拟：牛皮纸，水墨
AI-derived digital painting based on the painting (Left)
Simulated Material: Kraft paper, Ink
928px×1232px
2024

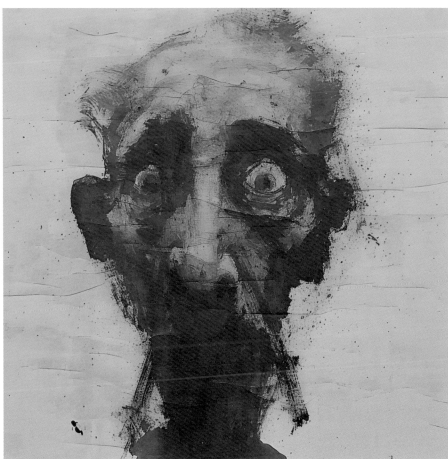

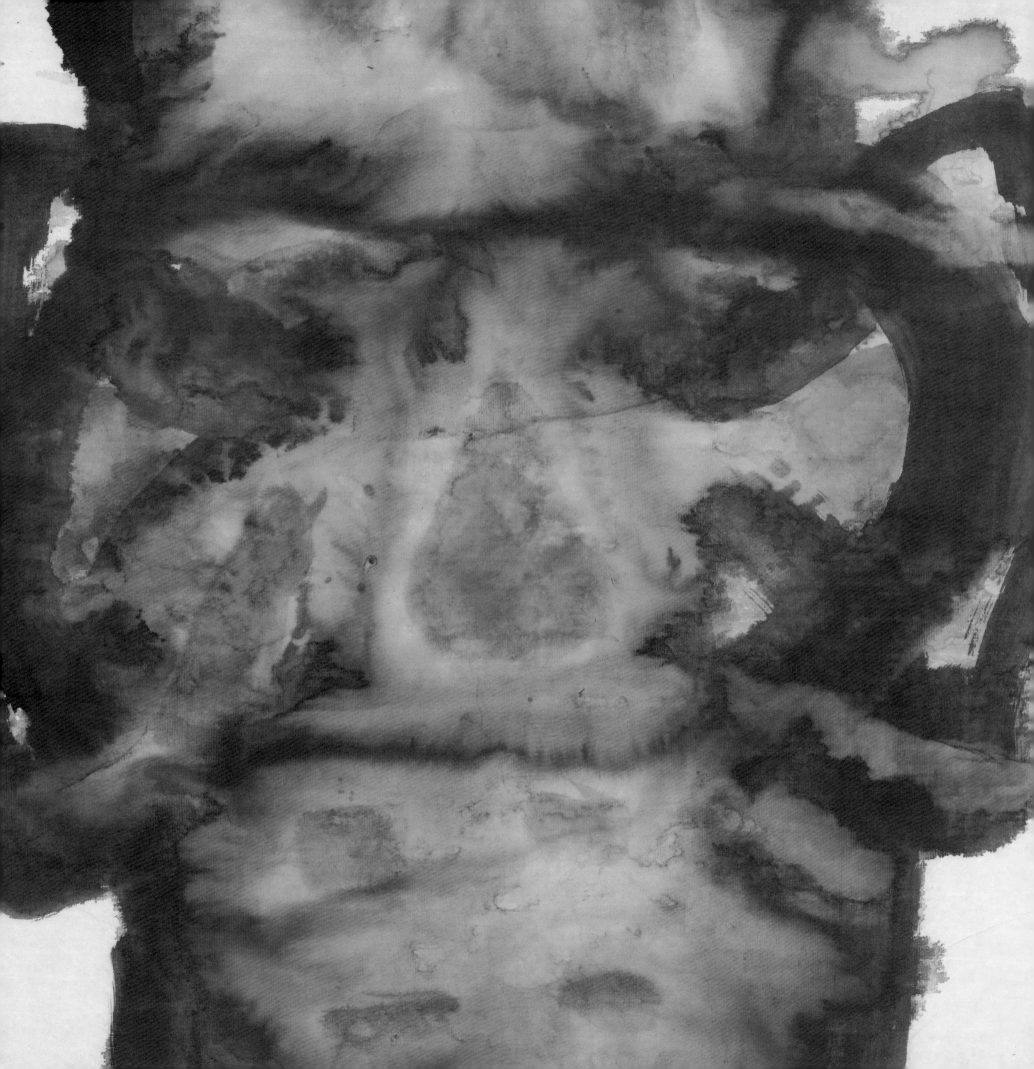

实验水墨
绘画材料：牛皮纸，水墨

Experimental Ink Wash
Painting
Artwork: Kraft paper, ink
55cm×60cm
2023

基于左图的 AI 衍生数字绘画
媒材模拟：牛皮纸，水墨

AI-derived digital painting based on the painting (Left)
Simulated Material: Kraft paper, Ink
928px × 1232px
2023

基于左图的 AI 衍生数字绘画
媒材模拟：牛皮纸，水墨

AI-derived digital painting based on the painting (Left)
Simulated Material: Kraft paper, Ink
928px × 1232px
2023

"我想对于创作者,

如果肉体走得过快,而灵魂没有跟上,

即作品只有夺目的奇观而缺乏精神的深度,

无法与观者的心灵产生共振,

这实则是一件很悲哀的事。"

"For creators,

if the body moves too fast while the soul fails to keep up,

their works are very likely to have dazzling spectacles but lack spiritual depth

and fail to resonate with the viewer's mind and soul,

which would be quite a tragic."

实验水墨
绘画材料：宣纸板，水墨

Experimental Ink Wash Painting
Artwork: Rice paper board, Ink
33cm × 46cm
2024

基于左图的 AI 衍生数字绘画
媒材模拟：宣纸板，水墨

AI-derived digital painting based on the painting (Left)
Simulated Material: Rice paper board, Ink
928px × 1232px
2024

基于左图的 AI 衍生数字绘画
媒材模拟：宣纸板，水墨

AI-derived digital painting based on the painting (Left)
Simulated Material: Rice paper board, Ink
928px × 1232px
2024

基于 226 页作品的 AI 衍生数字绘画
媒材模拟：宣纸板，水墨

AI-derived digital painting based on the painting (P226)
Simulated Material: Rice paper board, Ink
928px × 1232px
2024

基于 226 页作品的 AI 衍生数字绘画
媒材模拟：宣纸板，水墨

AI-derived digital painting based on the painting (P226)
Simulated Material: Rice paper board, Ink
928px × 1232px
2024

基于 226 页作品的 AI 衍生数字绘画
媒材模拟：宣纸板，水墨

AI-derived digital painting based on the painting
(Based on P226)
Simulated Material: Rice paper board, Ink
928px × 1232px
2024

基于 226 页作品的 AI 衍生数字绘画
媒材模拟：宣纸板，水墨

AI-derived digital painting based on the painting
(Based on P226)
Simulated Material: Rice paper board, Ink
928px × 1232px
2024

实验水墨
绘画材料：宣纸板，水墨

Experimental Ink Wash Painting
Artwork: Rice paper board, Ink
33cm × 46cm
2024

基于左图的 AI 衍生数字绘画
媒材模拟：宣纸板，水墨

AI-derived digital painting based on the painting (Left)
Simulated Material: Rice paper board, Ink
928px × 1232px
2024

基于左图的 AI 衍生数字绘画
媒材模拟：宣纸板，水墨

AI-derived digital painting based on the painting (Left)
Simulated Material: Rice paper board, Ink
928px × 1232px
2024

基于 232 页作品的 AI 衍生数字绘画
媒材模拟：宣纸板，水墨

AI-derived digital painting based on the painting (P232)
Simulated Material: Rice paper board, Ink
928px × 1232px
2024

基于 232 页作品的 AI 衍生数字绘画
媒材模拟：宣纸板，水墨

AI-derived digital painting based on the painting (P232)
Simulated Material: Rice paper board, Ink
928px × 1232px
2024

基于 232 页作品的 AI 衍生数字绘画　　　　　　　基于 232 页作品的 AI 衍生数字绘画
媒材模拟：宣纸板，水墨　　　　　　　　　　　　媒材模拟：宣纸板，水墨

AI-derived digital painting based on the painting (P232)　　AI-derived digital painting based on the painting (P232)
Simulated Material: Rice paper board, Ink　　　　　　　　Simulated Material: Rice paper board, Ink
928px × 1232px　　　　　　　　　　　　　　　　　　928px × 1232px
2024　　　　　　　　　　　　　　　　　　　　　　　2024

实验水墨
绘画材料：宣纸，水墨

Experimental Ink Wash Painting
Artwork：Rice paper, Ink
55cm × 60cm
2024

基于左图的 AI 衍生数字绘画
媒材模拟：宣纸，水墨

AI-derived digital painting based on the painting (Left)
Simulated Material：Rice paper, Ink
928px × 1232px
2024

基于左图的 AI 衍生数字绘画
媒材模拟：宣纸，水墨

AI-derived digital painting based on the painting (Left)
Simulated Material: Rice paper, Ink
928px × 1232px
2024

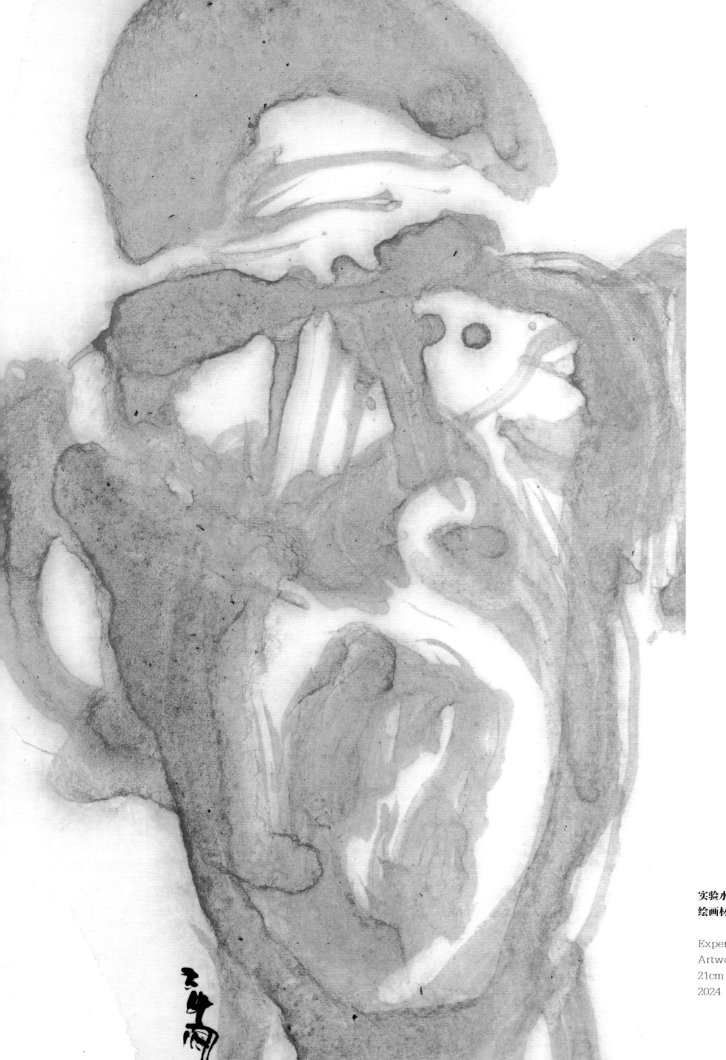

实验水墨
绘画材料：素描纸，水墨

Experimental Ink Wash Painting
Artwork:Sketch paper, Ink
21cm×28cm
2024

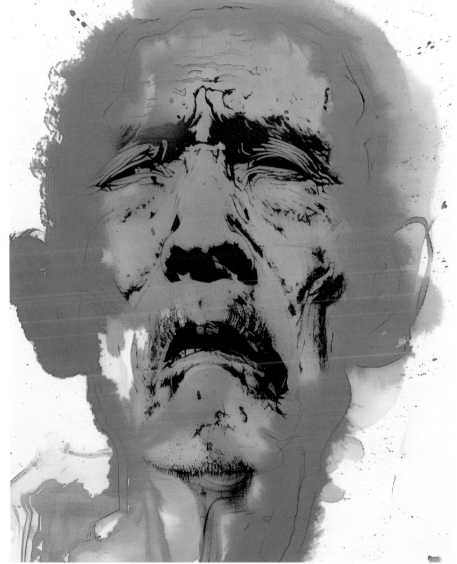

基于左图的 AI 衍生数字绘画
媒材模拟：素描纸，水墨

AI-derived digital painting based on the painting (Left)
Simulated Material:Sketch paper, Ink
928px × 1232px
2024

基于左图的 AI 衍生数字绘画
媒材模拟：素描纸，水墨

AI-derived digital painting based on the painting (Left)
Simulated Material:Sketch paper, Ink
928px × 1232px
2024

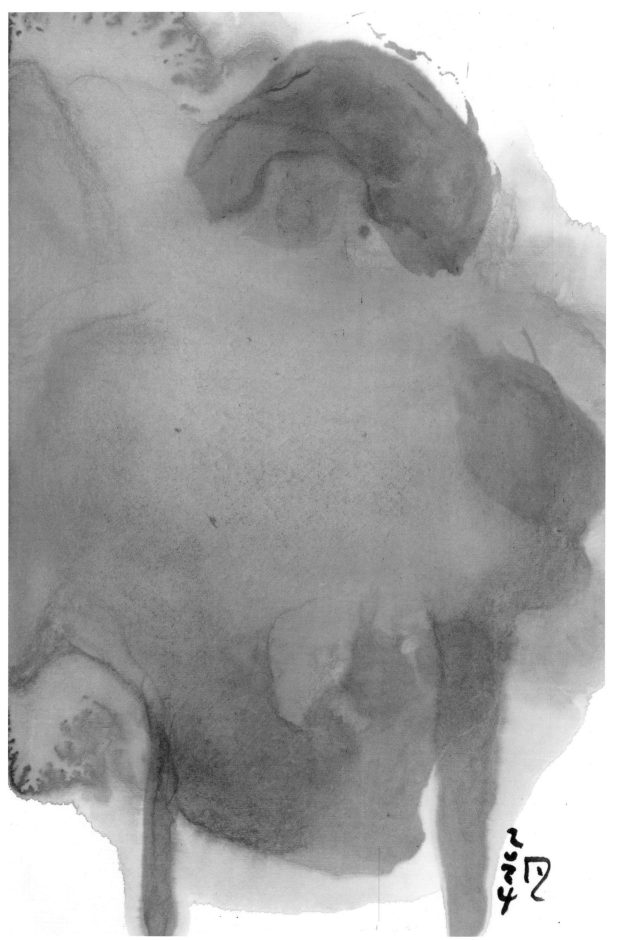

实验水墨
绘画材料：宣纸板，水墨

Experimental Ink Wash Painting
Artwork: Rice paper board, Ink
33cm × 46cm
2024

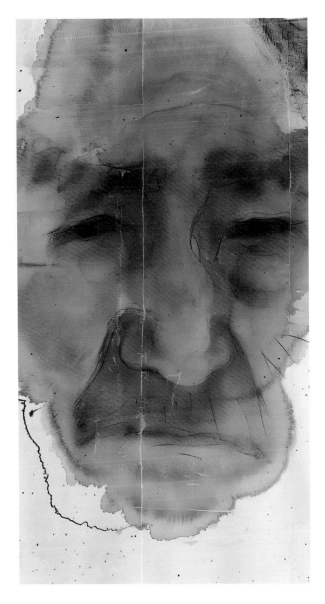
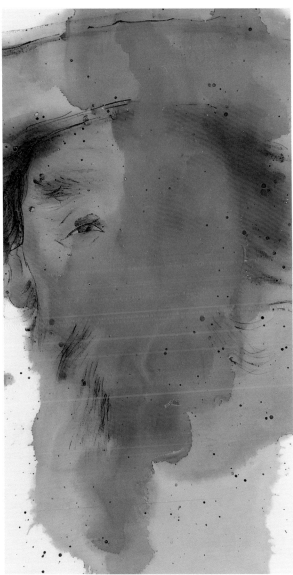

基于左图的 AI 衍生数字绘画
媒材模拟：宣纸板，水墨

AI-derived digital painting based on the painting (Left)
Simulated Material: Rice paper board, Ink
928px × 1232px
2024

基于左图的 AI 衍生数字绘画
媒材模拟：宣纸板，水墨

AI-derived digital painting based on the painting (Left)
Simulated Material: Rice paper board, Ink
928px × 1232px
2024

基于左图的 AI 衍生数字绘画
媒材模拟：宣纸板，水墨

AI-derived digital painting based on the painting (Left)
Simulated Material: Rice paper board, Ink
928px × 1232px
2024

实验水墨
绘画材料：素描纸，水墨

Experimental Ink Wash Painting
Artwork：Sketch paper, Ink
21cm×28cm
2024

基于左图的 AI 衍生数字绘画
媒材模拟：素描纸，水墨

AI-derived digital painting based on the painting (Left)
Simulated Material：Sketch paper, Ink
928px×1232px
2024

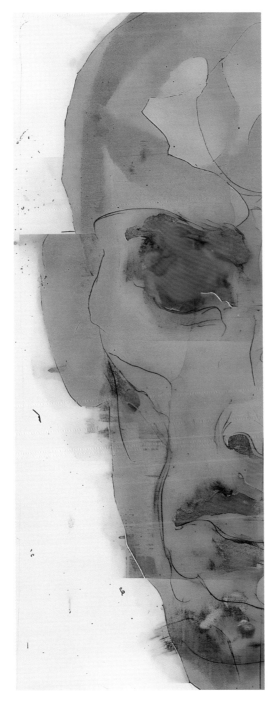

基于左图的 AI 衍生数字绘画
媒材模拟：素描纸，水墨

AI-derived digital painting based on the painting (Left)
Simulated Material: Sketch paper, Ink
928px × 1232px
2024

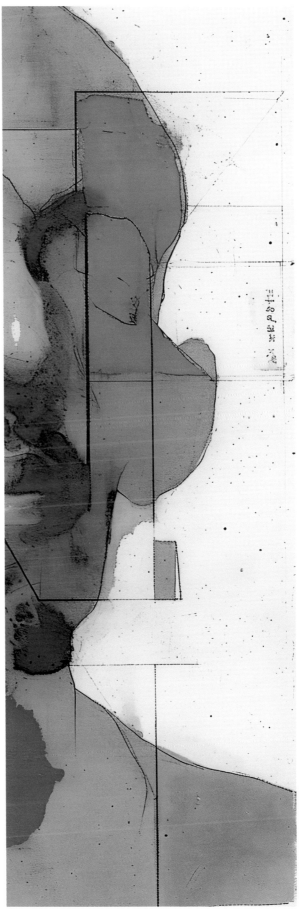

基于左图的 AI 衍生数字绘画
媒材模拟：素描纸，水墨

AI-derived digital painting based on the painting (Left)
Simulated Material: Sketch paper, Ink
928px × 1232px
2024

"实现通力合作,

某种意义上就是整合人类与人工智能在艺术创作领域的优势,

让二者的优势'做加法'。"

"Achieving collaboration essentially means

integrating the strengths of humans and artificial intelligence in the field of artistic creation,

allowing the advantages of both to complement each other."

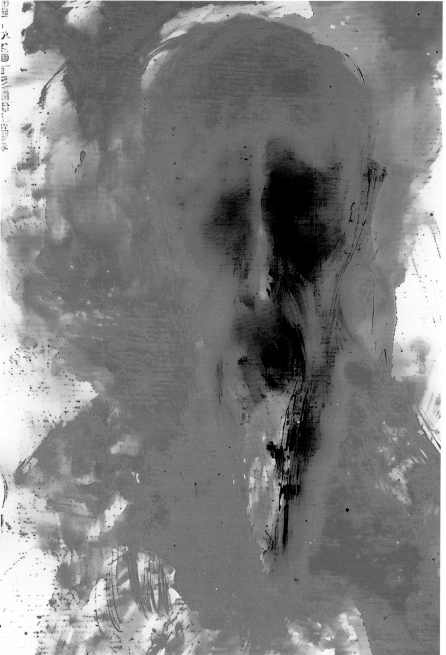

实验水墨
绘画材料：素描纸，水墨

Experimental Ink Wash Painting
Artwork: Sketch paper, Ink
21cm×28cm
2023

基于左图的 AI 衍生数字绘画
媒材模拟：素描纸，水墨

AI-derived digital painting based on the painting (Left)
Simulated Material: Sketch paper, Ink
928px×1232px
2023

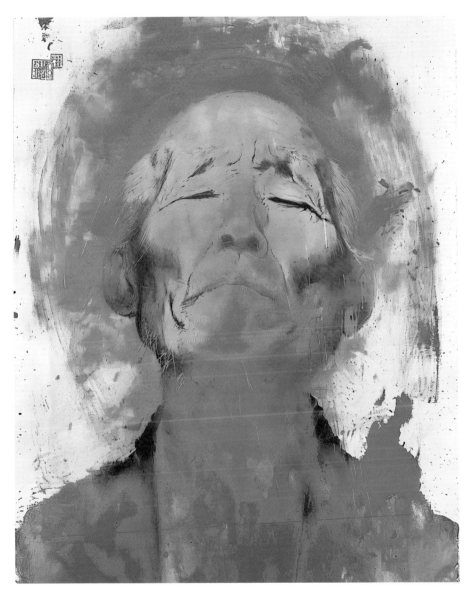

基于左图的 AI 衍生数字绘画
媒材模拟：素描纸，水墨

AI-derived digital painting based on the painting (Left)
Simulated Material: Sketch paper, Ink
928px × 1232px
2023

基于左图的 AI 衍生数字绘画
媒材模拟：素描纸，水墨

AI-derived digital painting based on the painting (Left)
Simulated Material: Sketch paper, Ink
928px × 1232px
2023

实验水墨
绘画材料：素描纸，水墨

Experimental Ink Wash Painting
Artwork: Sketch paper, Ink
21cm × 28cm
2023

基于左图的 AI 衍生数字绘画
媒材模拟：素描纸，水墨

AI-derived digital painting based on the painting (Left)
Simulated Material: Sketch paper, Ink
928px × 1232px
2023

基于左图的 AI 衍生数字绘画
媒材模拟：素描纸，水墨

AI-derived digital painting based on the painting (Left)
Simulated Material: Sketch paper, Ink
928px × 1232px
2023

实验水墨
绘画材料：素描纸，水墨

Experimental Ink Wash Painting
Artwork: Sketch paper, Ink
21cm × 28cm
2023

基于左图的 AI 衍生数字绘画
媒材模拟：素描纸，水墨

AI-derived digital painting based on the painting (Left)
Simulated Material: Sketch paper, Ink
1024px × 1024px
2023

基于左图的 AI 衍生数字绘画
媒材模拟：素描纸，水墨

AI-derived digital painting based on the painting (Left)
Simulated Material: Sketch paper, Ink
1024px × 1024px
2023

实验水墨
绘画材料：素描纸，水墨

Experimental Ink Wash Painting
Artwork: Sketch paper, Ink
21cm×28cm
2024

基于左图的 AI 衍生数字绘画
媒材模拟：素描纸，水墨

AI-derived digital painting based on the painting (Left)
Simulated Material:Sketch paper, Ink
928px × 1232px
2024

基于左图的 AI 衍生数字绘画
媒材模拟：素描纸，水墨

AI-derived digital painting based on the painting (Left)
Simulated Material:Sketch paper, Ink
928px × 1232px
2024

基于右图的 AI 衍生数字绘画
媒材模拟：素描纸，水墨

AI-derived digital painting based on the painting (Right)
Simulated Material：Sketch paper, Ink
928px × 1232px
2023

实验水墨
绘画材料：素描纸，水墨

Experimental Ink Wash Painting
Artwork：Sketch paper, Ink
21cm × 28cm
2023

基于左图的 AI 衍生数字绘画
媒材模拟：素描纸，水墨

AI-derived digital painting based on the painting (Left)
Simulated Material：Sketch paper, Ink
928px × 1232px
2023

实验水墨
绘画材料：素描纸，水墨

Experimental Ink Wash
Painting
Artwork: Sketch paper, Ink
33cm × 46cm
2022

基于左图的 AI 衍生数字绘画
媒材模拟：素描纸，水墨

AI-derived digital painting
based on the painting (Left)
Simulated Material: Sketch
paper, Ink
928px × 1232px
2023

第四章
发现：碰撞数字绘画艺术的可能性
Chapter 4
Discovery: Exploring the Possibilities of Digital Painting Art and AI

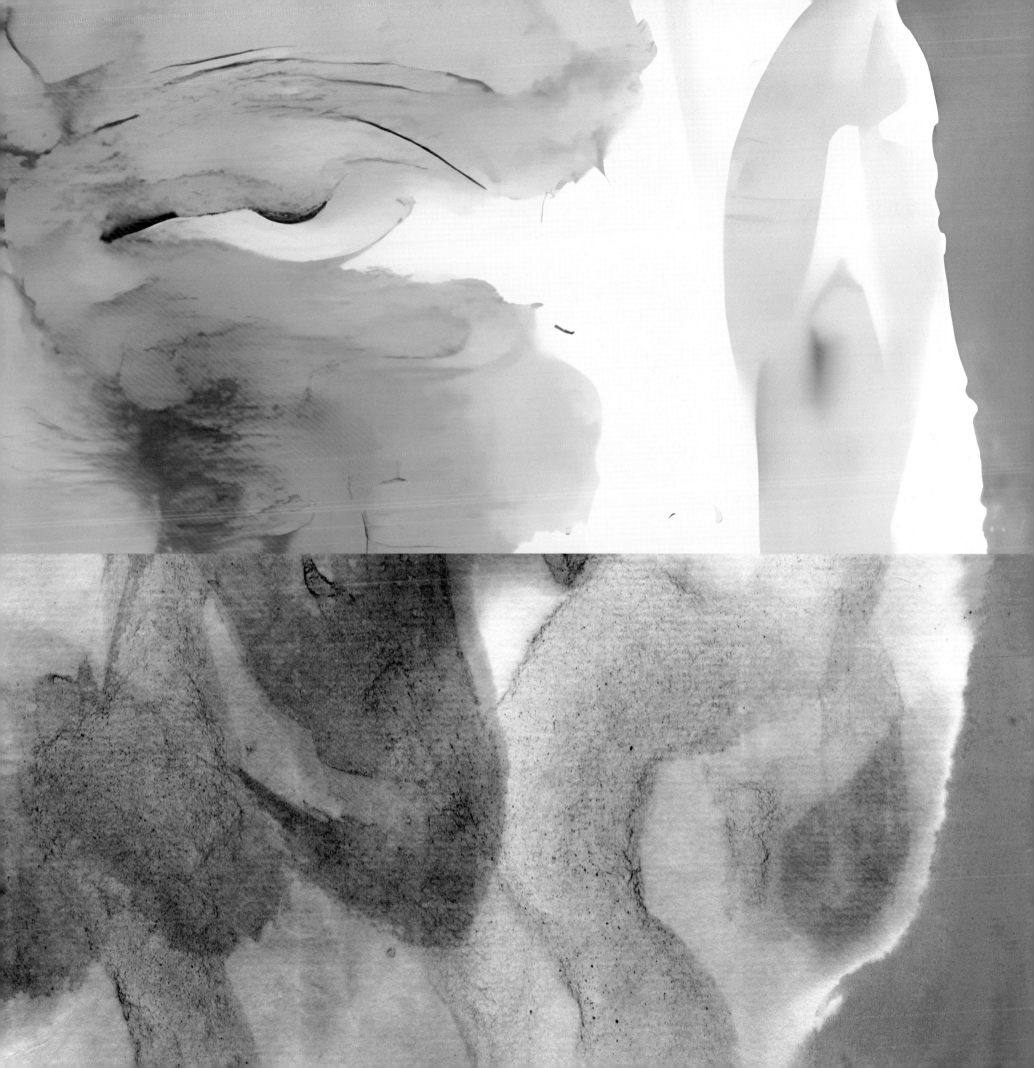

采访10：相较于《墨像》一书中的黑白水墨，本书中收录了一些彩色水墨作品。

凡： 我们知道传统意义的水墨画讲究的都是"墨分五彩""浓淡干湿"，墨并不仅仅是一种颜色，而是通过调和水与墨的配比，加之用笔的轻重缓急、皴擦点染等手法，产生颜色的浓淡之分与墨的干湿之分，因而在一幅传统水墨绘画中，即使只用单一的墨色，也可以使画面产生丰富的层次变化，在浓淡虚实的变化间赋予自然物象以抽象的意蕴。本书中有意加入很多黑白灰调之外的色彩，主要还是考虑到方便人工智能系统识别的需求。目前人工智能对黑白灰色阶的识别还不够精准，加之大笔阔写的笔法与造型风格的抽象，我发现当我们向系统输入黑白水墨作品时，常常会出现构图、造型、用线上的"误读"，输入彩色水墨作品后，我发现它对图像理解的精准度相应有了提升。也正是因为适当加入了一些彩色，我个人绘制这些作品时的体验感也有了相应的丰富，也希望读者能够获得更为丰富的审美体验，欢迎大家一同来辨别这些画作中哪些由人类艺术家创作，而哪些又是通过人工智能技术生成的。

Chapter 4 Discovery: Exploring the Possibilities of Digital Painting Art and AI

实验水墨
绘画材料：宣纸板，水墨，水彩

Experimental Ink Wash Painting
Artwork: Rice paper board, Ink, Watercolor
33cm × 46cm
2023

第四章 发现：碰撞数字绘画艺术的可能性

实验水墨
绘画材料：宣纸，水墨、水彩

Experimental Ink Wash Painting
Artwork：Rice paper, Ink, Watercolor
55cm×60cm
2023

Q10: Compared to the black and white ink paintings in the previous volume *Faces in Ink*, this volume seems to have included more colored ink works.

Fan: We know that traditional ink painting emphasizes "five colors" which refers to coke, rich, heavy, light, clear, or thick, light, dry, wet and black and "variations in ink density and wetness". Ink is not just a color, but through the blending of water and ink, combined with techniques such as the weight and urgency of strokes, and the rubbing and dabbing of the brush, it creates variations in color intensity and ink dryness. Thus, in a traditional ink painting, even using a single ink color can create rich changes in layers, imbuing natural objects with abstract meanings within the variations of intensity and reality. In this book, I intentionally added many colors beyond black, white, and gray tones, mainly considering the needs of artificial intelligence systems to analyze the work as data more precisely. Currently, artificial intelligence is not accurate enough in recognizing black, white, and gray tones, especially when the gradual change is not distinct enough. Moreover, with the abstract style of broad-brush writing and modeling, I found that when we input black and white ink works into the system, there are often "misinterpretations" in composition, modeling, and line usage. After inputting colored ink works, its accuracy in understanding images had correspondingly improved a lot. Moreover, it's precisely because of the addition of some colors that my personal experience in creating these works has accordingly become richer. I also hope that readers can gain a more diverse aesthetic experience. I welcome everyone to discern which artworks are created by human artists and which ones are generated through artificial intelligence technology.

第四章 发现：碰撞数字绘画艺术的可能性

实验水墨 基于左图的 AI 衍生数字绘画
绘画材料：素描纸，水墨 媒材模拟：素描纸，水墨

Experimental Ink Wash Painting AI-derived digital painting based on the painting (Left)
Artwork: Sketch paper, Ink Simulated Material: Sketch paper, Ink
21cm×28cm 928px×1232px
2023 2023

采访 11：在本书中，您尝试让人工智能学习您的水墨绘画作品，以您的基准看，在向算法输入海量作品后，它的"学习效果"如何？

凡： 我提供给人工智能的作品在技法与媒材应用上都具有鲜明的中国民族性，如"散点透视""墨分五彩"等，但也在构图、造型、空间、视点中融入了一定的西方美术思维。在十多位同学的辅助下，我们将这些绘画作品扫描成图，建立了初具规模的图像数据库。"基于中国审美的图像数据库，人工智能究竟能识别出什么，又能生成出什么"是我在本书中最想要和读者探讨的核心。当我看到人工智能生成的图像时，老实说第一反应是忍俊不禁。因为很明显的是，它生成的图像"并不那么中国"——它对图片的理解与图片的生成机制显然还是遵循着西方的视觉思维与审美范式，追求构建某种"具象"的"形象"而非"抽象"的"神韵"与"意境"，它看似模仿了数据库的笔法，但却无法复刻数据库中图像所传达的"临场感"与"情境感"，也正是基于这个发现，我突然萌生了"与人工智能赛跑"的想法，并进一步意识到在作品中传达中国审美的必要性。

第四章　发现：碰撞数字绘画艺术的可能性

基于 264 页作品的 AI 衍生数字绘画
媒材模拟：素描纸，水墨

AI-derived digital painting based on the painting (P264)
Simulated Material: Sketch paper, Ink
928px × 1232px
2023

基于 264 页作品的 AI 衍生数字绘画
媒材模拟：素描纸，水墨

AI-derived digital painting based on the painting (P264)
Simulated Material: Sketch paper, Ink
928px × 1232px
2023

Q11: In this volume, you attempted to let artificial intelligence learn from your ink painting works. Based on your standards, after inputting a massive amount of works into the algorithm, how would you evaluate the outcome?

Fan: The works I provided to artificial intelligence have a distinct Chinese national style in terms of techniques and media application, such as "cavalier perspective" and "five colors" which refers to coke, rich, heavy, light, clear, or thick, light, dry, wet and black, but also incorporate certain Western artistic methods in composition, modeling, space, and perspective. With the assistance of more than ten classmates, we scanned these paintings and established a preliminary image database. "What artificial intelligence can recognize and generate based on a database of Chinese aesthetics" is the core issue I most want to discuss with readers in this book. When I saw the images generated by artificial intelligence, to be honest, I bursted out laughing. Because it was quite obvious that the images it generated were "not very Chinese" — its understanding of the images and the mechanism of generated images evidently still followed Western visual patterns and aesthetic paradigms, pursuing the construction of a certain "concrete" "image" rather than the "abstract" "charm" and "artistic conception." It seemed to imitate the brushstrokes in the database, but it could not replicate the "sense of immediacy" and "sense of virtual realm" conveyed by the images in the database. It was precisely based on this discovery that I suddenly had the idea of "racing against artificial intelligence" and further realized the necessity of conveying Chinese aesthetics in the works.

采访 12：可不可以这样理解，生成式的人工智能绘画当前的局限很大程度上在于审美范式的局限？

凡： 这是一个方面，其根本很大程度上在于当前人工智能大数据的采集还是以西方国家流行文化或称主流文化的产出为主，因此在绘画生成的机制上鲜少遵循中国民族化的视觉思维，这点无疑是令人兴奋的。但生成式人工智能绘画的局限在此之外，还有人工智能对自然的体认和对"境界""心灵""情绪"等抽象概念的理解欠缺有关。这也是我一再强调艺术创作中"即兴"之美的原因所在。我希望在作品中唤起观者的某种心理体验，有时候你可能在这个形象当中看到一种呐喊，看到一种悲哀，看到一种痛苦，看到一种快乐（好像我并不怎么表达快乐），那种瞬间产生的、发乎于日常体验的情绪往往极具张力，观者在这一张力的震慑下，其自身的经验与想象被串联于激发，愉悦感随之油然而生……而人工智能是无法理解这种抽象的情感的，它只能复刻具体的、既定的形象，但情感是流动的、变化的，也正是因此，人类艺术家体察生活本质的能力在人工智能时代变得尤为重要。

Chapter 4 Discovery: Exploring the Possibilities of Digital Painting Art and AI

实验水墨
绘画材料：餐巾纸，水墨，茶水
Experimental Ink Wash Painting
Artwork:Napkin paper, Ink, Tea
12cm × 14cm
2023

实验水墨
绘画材料：信封纸，水墨

Experimental Ink Wash Painting
Artwork：Envelope paper, Ink
21cm×28cm
2022

Q12: So when it comes to the current limitations of generative artificial intelligence painting, it's more about the aesthetic paradigms, isn't it?

Fan: This is one aspect. Fundamentally, the current composition of big data used by artificial intelligence is mainly based on the output of Western countries' popular culture or mainstream culture. Therefore, there is a lack of adherence to Chinese-nationalized visual tradition in the mechanism of AI generated art, which is undoubtedly exciting. However, beyond the limitations of artificial intelligence generated painting, there is also a lack of artificial intelligence's recognition of nature and its understanding of abstract concepts such as "realm", "mind", and "emotion". This is also the reason why I repeatedly emphasize the beauty of "improvisation" in artistic creation. I hope to evoke a certain psychological experience in the work. Sometimes you can see a kind of abreaction, sorrow, pain, or joy (honestly I don't express joy much) in this image. The momentary emotions arising from daily experiences often have great tension. Under the impact of this tension, the viewer's own experiences and imagination are stimulated, and a sense of pleasure emerges... However, artificial intelligence cannot understand these abstract emotions. It can only replicate concrete, established images, but emotions are fluid and changing. Therefore, the ability of human artists to perceive the essence of life becomes particularly important in the era of artificial intelligence.

采访13：您认为未来人类和人工智能将如何以数字绘画的方式"讲好中国故事"？

凡： 首先我认为人工智能时代带有某种"不可逆"的属性。技术的发展永远是向前的，正所谓"物竞天择，适者生存"，我们与其悲叹"技术吞噬人类文明"，不如思考人类无法被技术取代的优势是什么，以及人类的优势如何与人工智能实现优势互补。人类的优势在于对生活细致入微的感知力与兼容理性与感性的创造力，而人工智能的优势则在于大数据的综合分析能力，基于某个概念、某种风格、某类题材给出多元化的视觉阐释路径，极大程度上扩展个体的艺术与人文视野，让个体的创作活动更具有群体性。落实到"讲好中国故事"这点，我想正如中国动画学派一贯强调的"不重复自己，不模仿他人"一般，我们应当在日新月异的技术与人文环境下增强自身的原创力，赋予民族化的工艺技法与美学风格以令人耳目一新的形态，在创作中融入艺术家对当代国人火热生活图景的感知，在"已知"中创造"未知"，传达当代国人的情感诉求与未来期许，由此来讲好真正富有"精、气、神"、诉诸心灵的"中国故事"。

Chapter 4　Discovery: Exploring the Possibilities of Digital Painting Art and AI

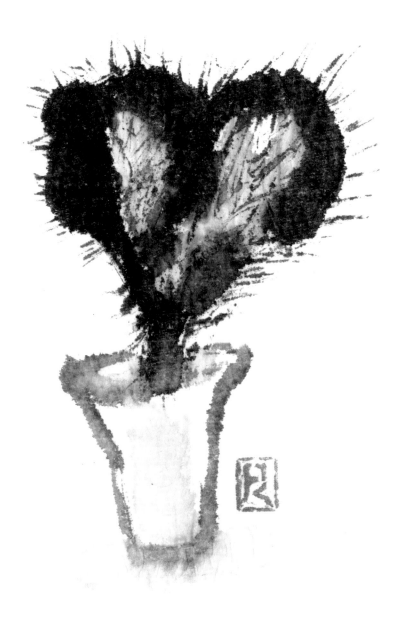

实验水墨
绘画材料：餐巾纸，水墨，茶水

Experimental Ink Wash Painting
Artwork:Napkin paper, Ink, Tea
12cm×14cm
2023

Q13: How do you think humans and artificial intelligence can better tell Chinese stories through digital painting in the future?

Fan: Firstly, I believe that the era of artificial intelligence carries a certain "irreversible" attribute. Technological development always moves forward, as the saying goes, "survival of the fittest". Instead of lamenting that "technology is devouring human civilization", we should consider what advantages humans have that cannot be replaced by technology and how human advantages can complement those of artificial intelligence. The advantages of humans lie in their meticulous perception of life and their creative ability that encompasses both rationality and emotion. On the other hand, the advantage of artificial intelligence lies in its comprehensive analysis ability of big data, which provides diverse visual establishments based on certain concepts, styles, or themes, greatly expanding the individual's artistic and humanistic perspective and making individual creative activities more collective. In terms of telling Chinese stories, I believe that, just as the Chinese Animation School consistently emphasizes "Not repeating oneself, not imitating others", we should enhance our originality in the rapidly changing technological and cultural environment. We should infuse nationalized craft techniques and aesthetic styles with refreshing forms, integrate the artist's perception of the spectacles of life that belongs to contemporary Chinese people into our creations, bringing alive "unknowns" within the "known", and convey the emotional appeals and future expectations of contemporary Chinese people. Through this, we can truly tell "Chinese stories" that are rich in "essence, vitality, and spirit", and also appealing to the soul.

Chapter 4　Discovery: Exploring the Possibilities of Digital Painting Art and AI

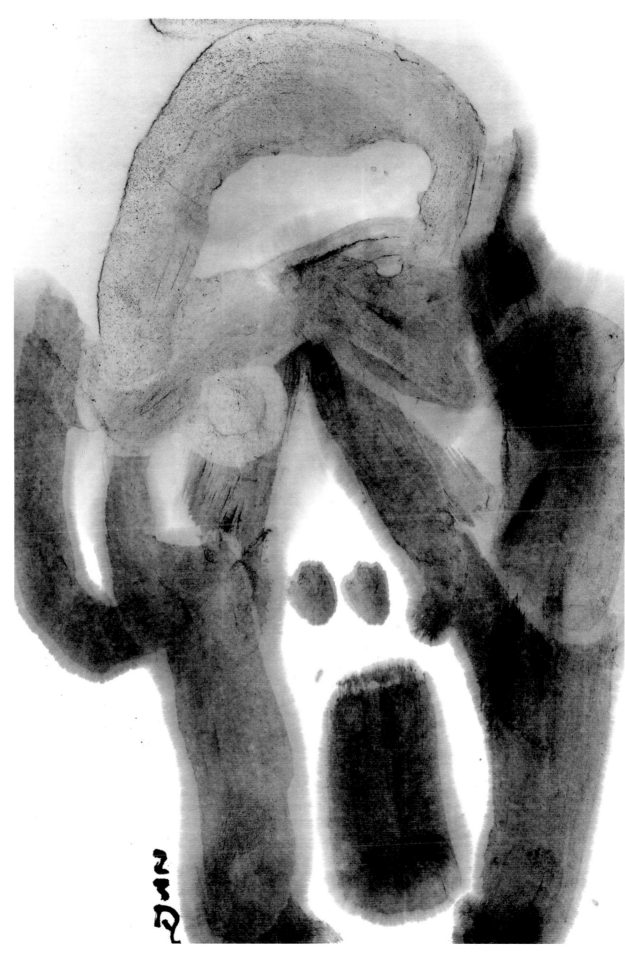

实验水墨
绘画材料：宣纸板，水墨

Experimental Ink Wash Painting
Artwork: Rice paper board, Ink
33cm × 46cm
2023

实验水墨
绘画材料：素描纸，水墨

Experimental Ink Wash Painting
Artwork: Sketch paper, Ink
21cm × 28cm
2023

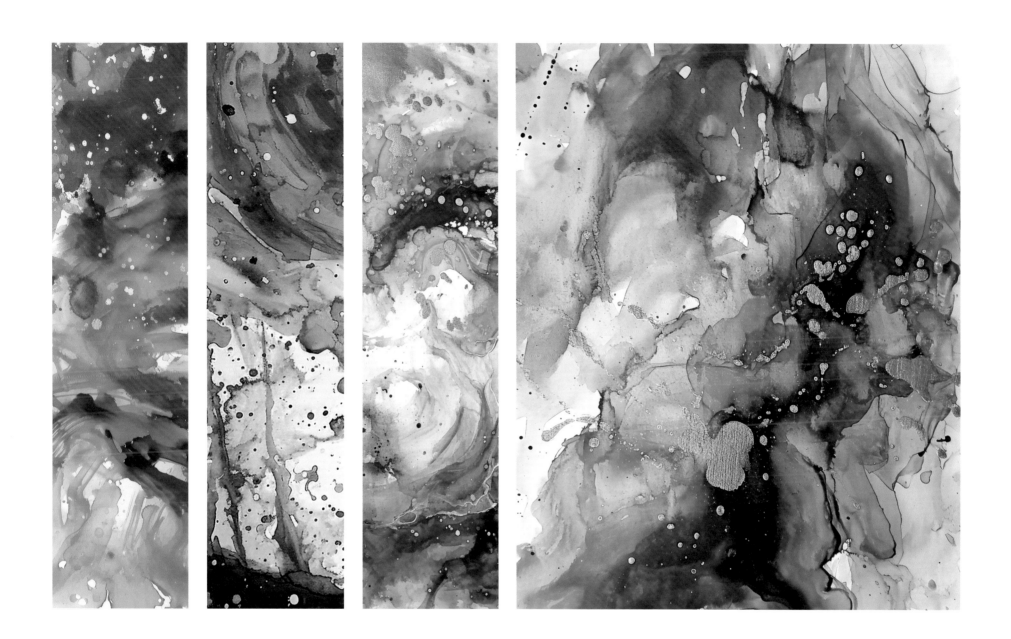

基于左图的 AI 衍生数字绘画
媒材模拟：素描纸，水墨

AI-derived digital painting based on the painting (Left)
Simulated Material: Sketch paper, Ink
928px × 1232px
2023

基于 276 页作品的 AI 衍生数字绘画
媒材模拟：素描纸，水墨

AI-derived digital painting based on the painting (P276)
Simulated Material: Sketch paper, Ink
928px × 1232px
2023

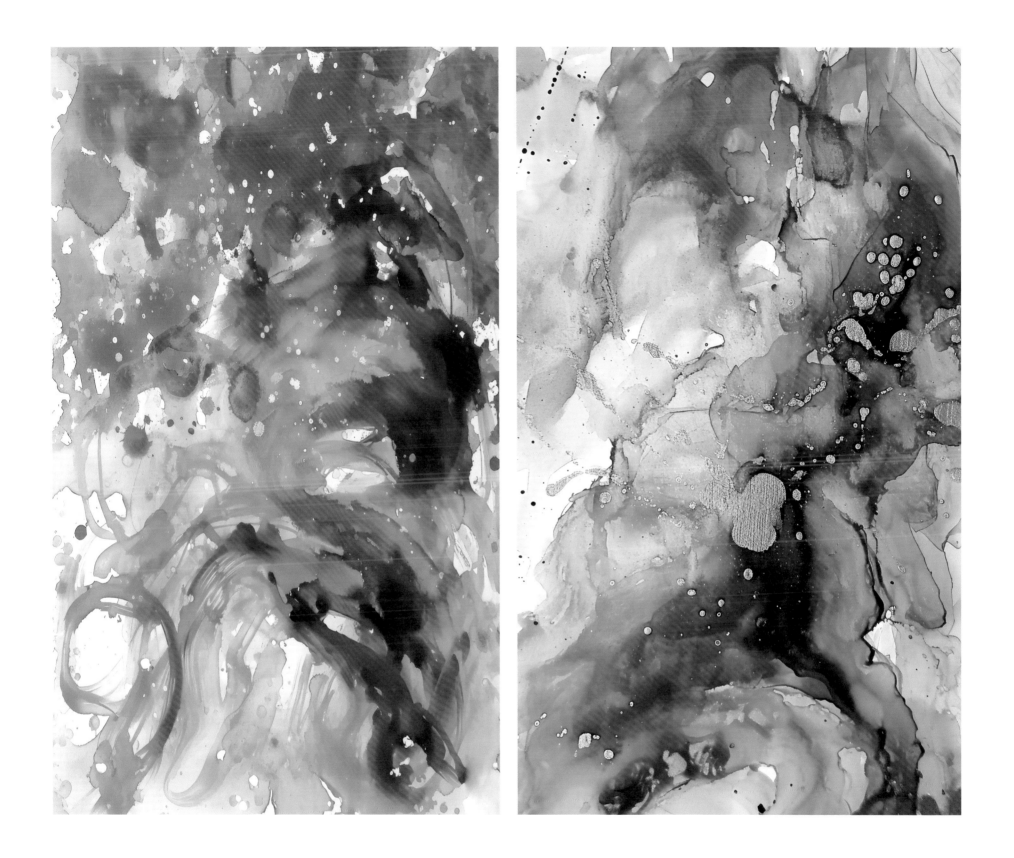

基于 276 页作品的 AI 衍生数字绘画
媒材模拟：素描纸，水墨

AI-derived digital painting based on the painting (P276)
Simulated Material: Sketch paper, Ink
928px × 1232px
2023

实验水墨
绘画材料：宣纸板，颜料，水墨

Experimental Ink Wash Painting
Artwork:Rice paper board, Pigment, Ink
33cm × 46cm
2024

实验水墨
绘画材料：宣纸，水墨

Experimental Ink Wash Painting
Artwork:Rice paper, Ink
55cm × 60cm
2023

实验水墨
绘画材料：宣纸板，水彩

Experimental Ink Wash Painting
Artwork:Rice paper board, Watercolor
33cm×46cm
2023

实验水墨
绘画材料：宣纸板，水彩

Experimental Ink Wash Painting
Artwork:Rice paper board, Watercolor
33cm×46cm
2023

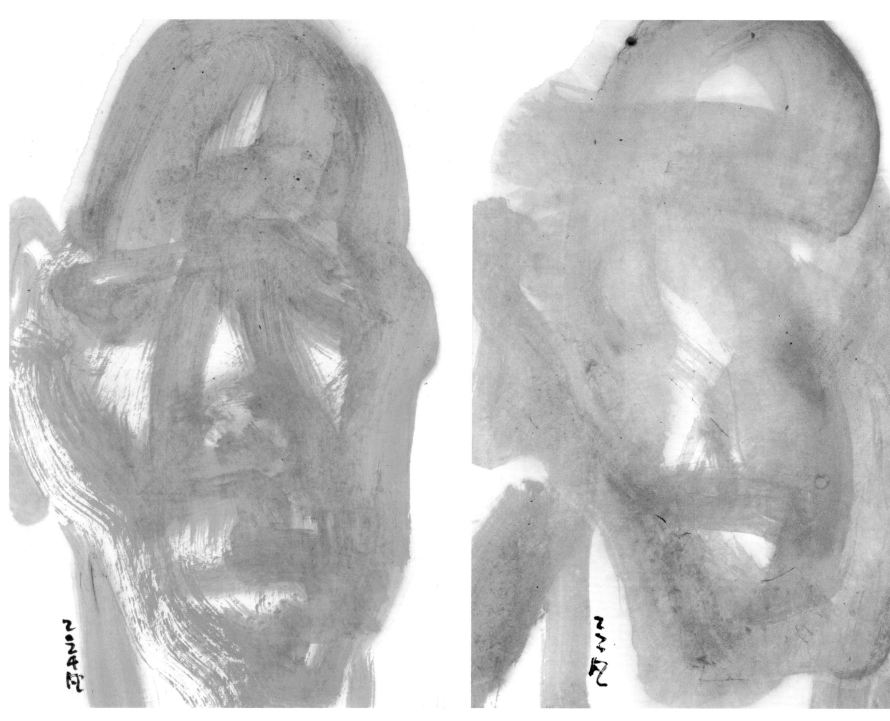

实验水墨
绘画材料：素描纸，水墨

Experimental Ink Wash Painting
Artwork：Sketch paper, Ink
21cm × 28cm
2024

实验水墨
绘画材料：素描纸，水墨

Experimental Ink Wash Painting
Artwork：Sketch paper, Ink
21cm × 28cm
2023

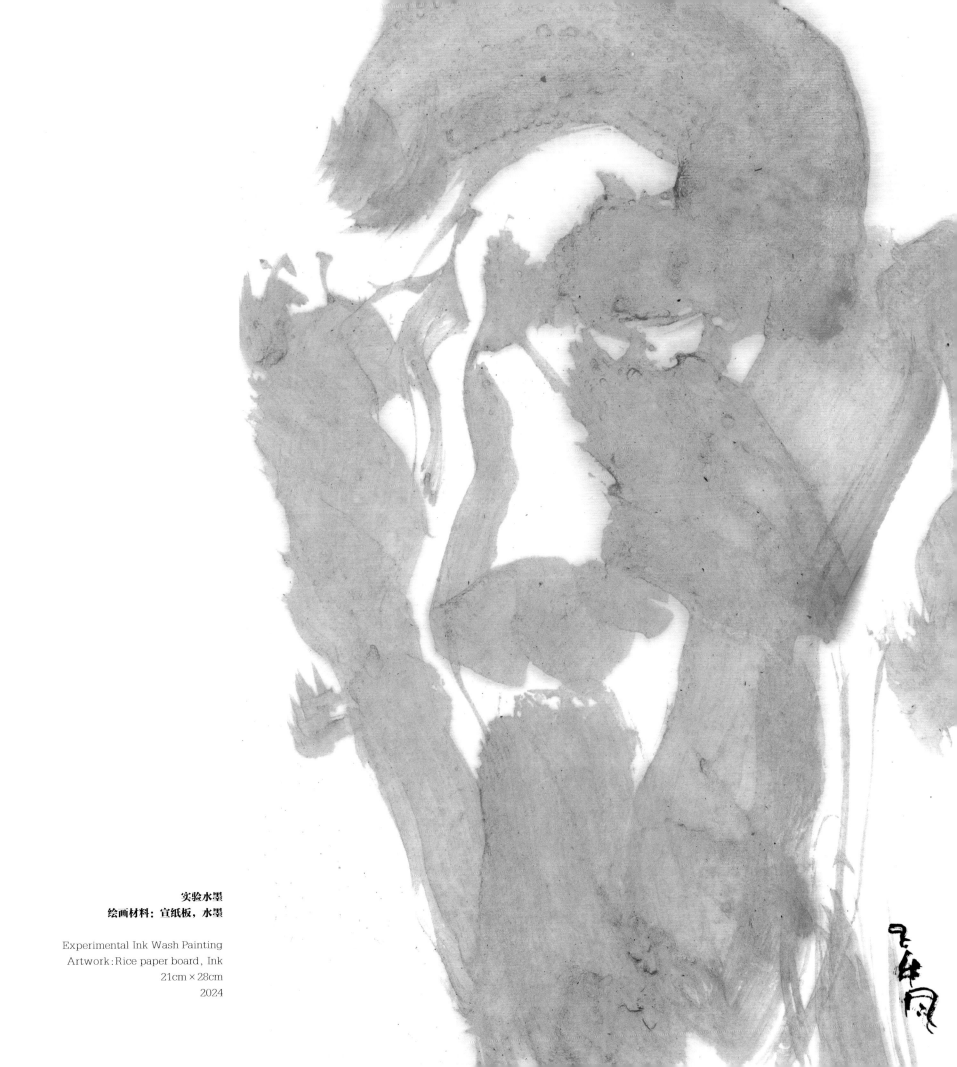

实验水墨
绘画材料：宣纸板，水墨

Experimental Ink Wash Painting
Artwork: Rice paper board, Ink
21cm × 28cm
2024

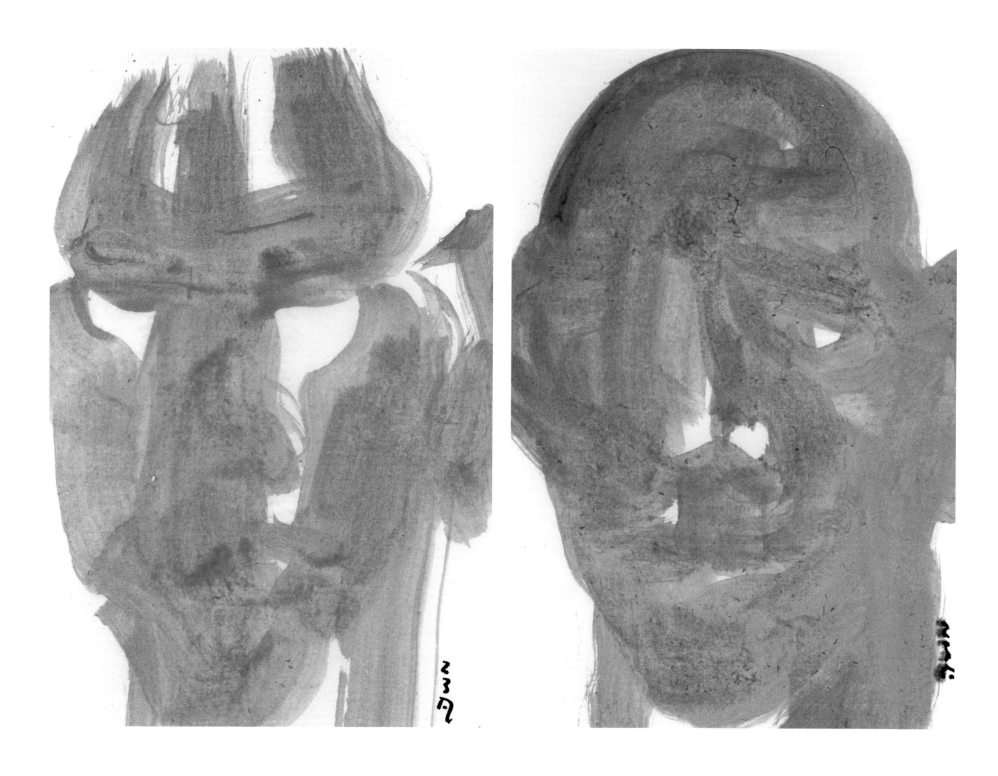

实验水墨
绘画材料：宣纸板，水彩

Experimental Ink Wash Painting
Artwork：Rice paper board, Watercolor
33cm×46cm
2023

实验水墨
绘画材料：宣纸板，水彩

Experimental Ink Wash Painting
Artwork：Rice paper board, Watercolor
33cm×46cm
2023

实验水墨
绘画材料：宣纸板，彩色颜料

Experimental Ink Wash Painting
Artwork：Rice paper board, Coloured pigment
33cm×46cm
2023

实验水墨
绘画材料：宣纸板，彩色颜料

Experimental Ink Wash Painting
Artwork：Rice paper board, Coloured pigment
33cm×46cm
2023

实验水墨
绘画材料：宣纸板，水墨，颜料

Experimental Ink Wash Painting
Artwork: Rice paper board, Ink, Pigment
33cm × 46cm
2023

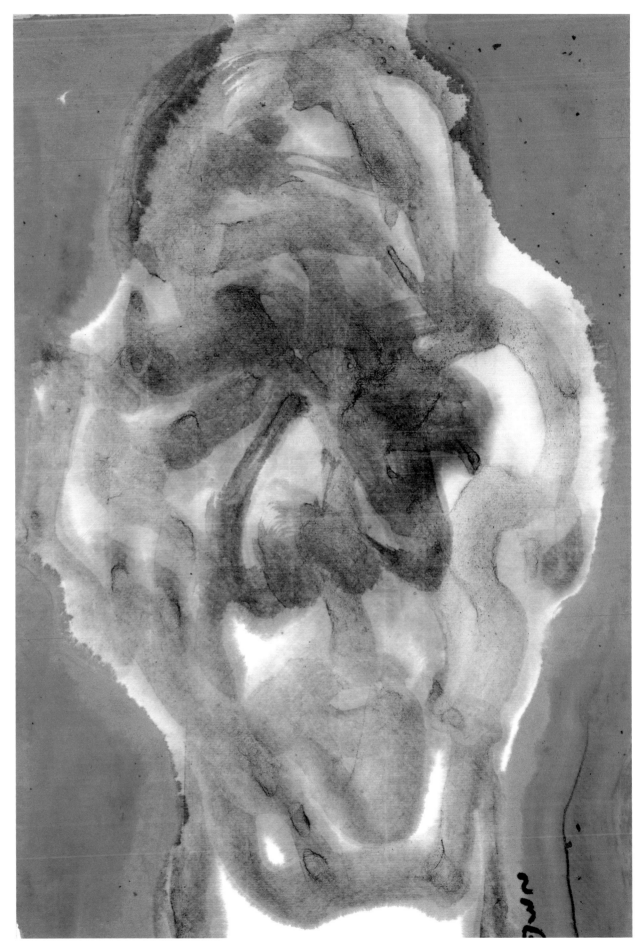

实验水墨
绘画材料：宣纸板，水墨，颜料

Experimental Ink Wash Painting
Artwork: Rice paper board, Ink, Pigment
33cm × 46cm
2023

"'基于中国审美的图像数据库，

人工智能究竟能识别出什么，又能生成出什么'

是我在本书中最想要和读者探讨的内容。"

"'What artificial intelligence can recognize and generate based on a database of Chinese aesthetics'

is the core issue I most want to discuss with readers in this book."

AI 绘画：当代水墨艺术的"正发生" | 289

实验水墨
绘画材料：宣纸板，颜料

Experimental Ink Wash Painting
Artwork: Rice paper board, Pigment
33cm × 46cm
2023

实验水墨
绘画材料：宣纸板，颜料

Experimental Ink Wash Painting
Artwork: Rice paper board, Pigment
33cm × 46cm
2023

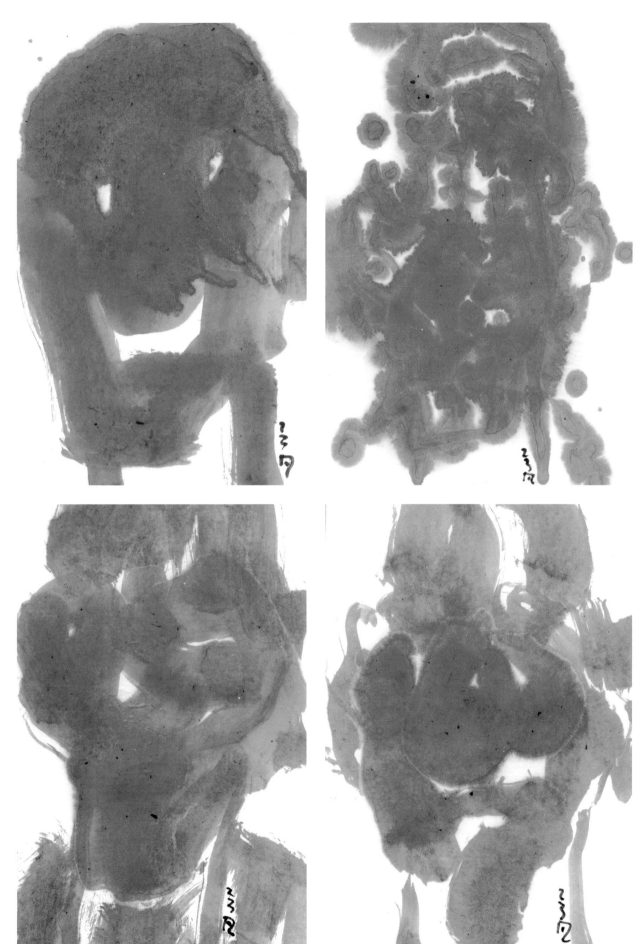

实验水墨
绘画材料：宣纸板，颜料

Experimental Ink Wash Painting
Artwork：Rice paper board, Pigment
33cm×46cm
2023

实验水墨
绘画材料：宣纸板，颜料

Experimental Ink Wash Painting
Artwork：Rice paper board, Pigment
33cm×46cm
2023

实验水墨
绘画材料：牛皮纸，水墨
Experimental Ink Wash Painting
Artwork：Kraft paper, Ink
55cm×60cm
2023

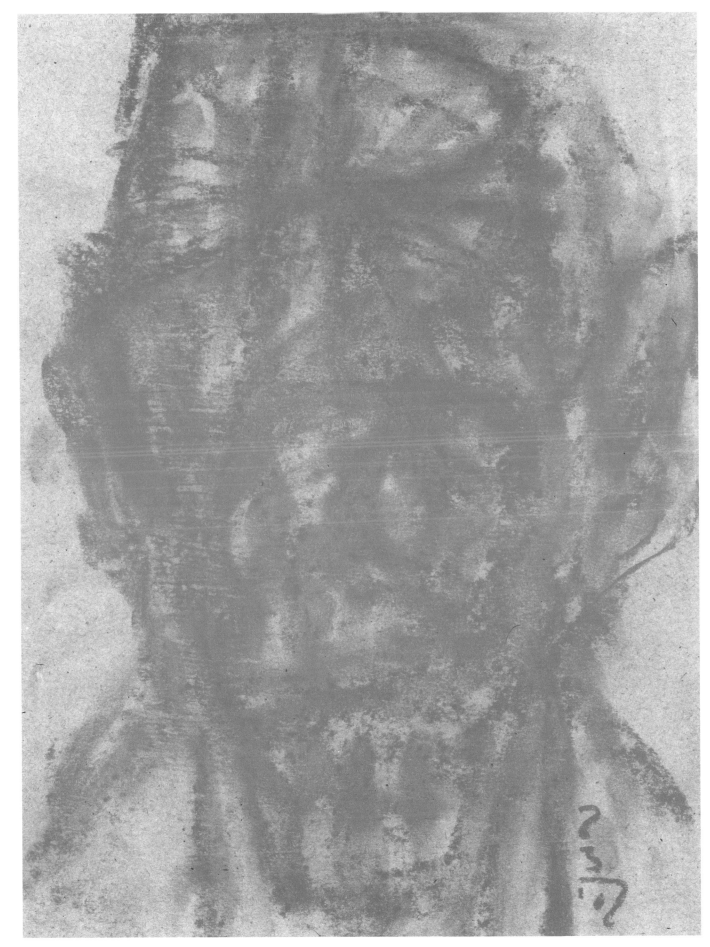

实验水墨
绘画材料：牛皮纸，色粉

Experimental Ink Wash Painting
Artwork: Kraft paper, Pastel
33cm × 46cm
2023

实验水墨
绘画材料：宣纸板，颜料

Experimental Ink Wash Painting
Artwork：Rice paper board, Pigment
33cm × 46cm
2023

实验水墨
绘画材料：宣纸板，颜料

Experimental Ink Wash Painting
Artwork: Rice paper board, Pigment
33cm × 46cm
2023

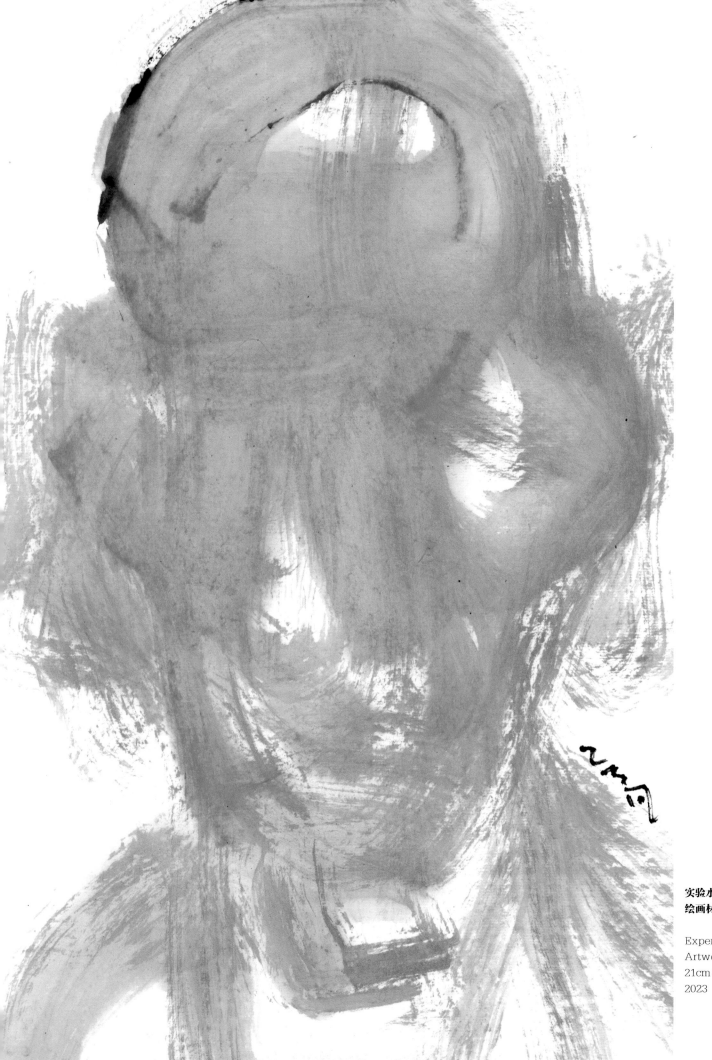

实验水墨
绘画材料：宣纸板，水墨

Experimental Ink Wash Painting
Artwork: Rice paper board, Ink
21cm × 28cm
2023

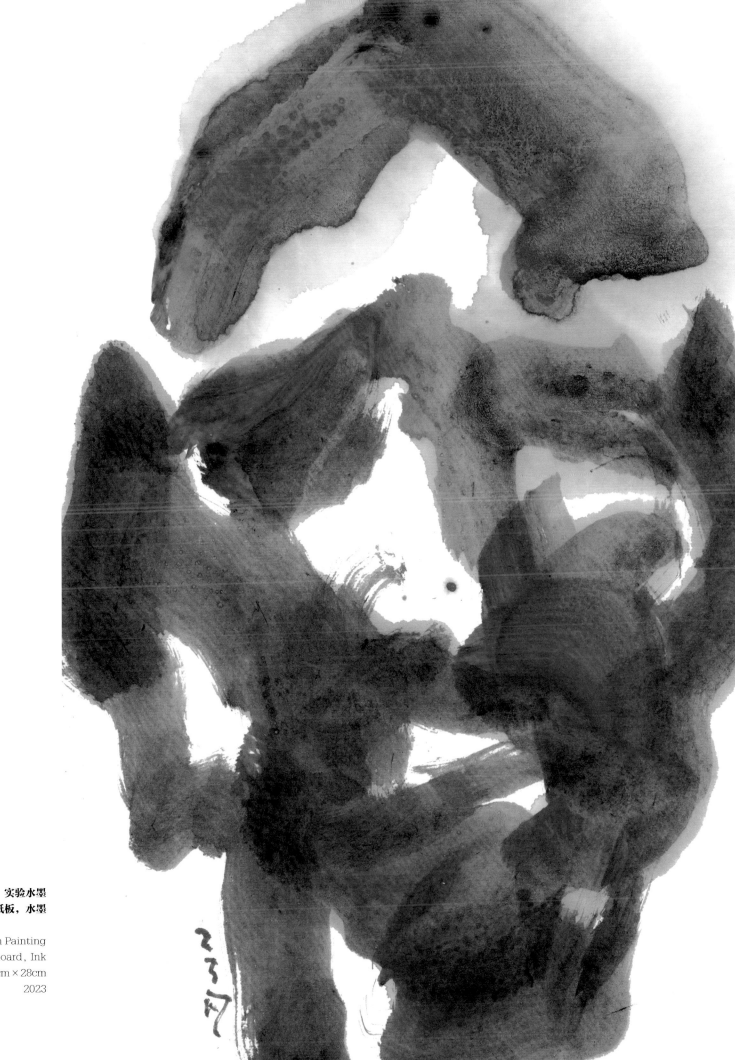

实验水墨
绘画材料：宣纸板，水墨

Experimental Ink Wash Painting
Artwork：Rice paper board, Ink
21cm×28cm
2023

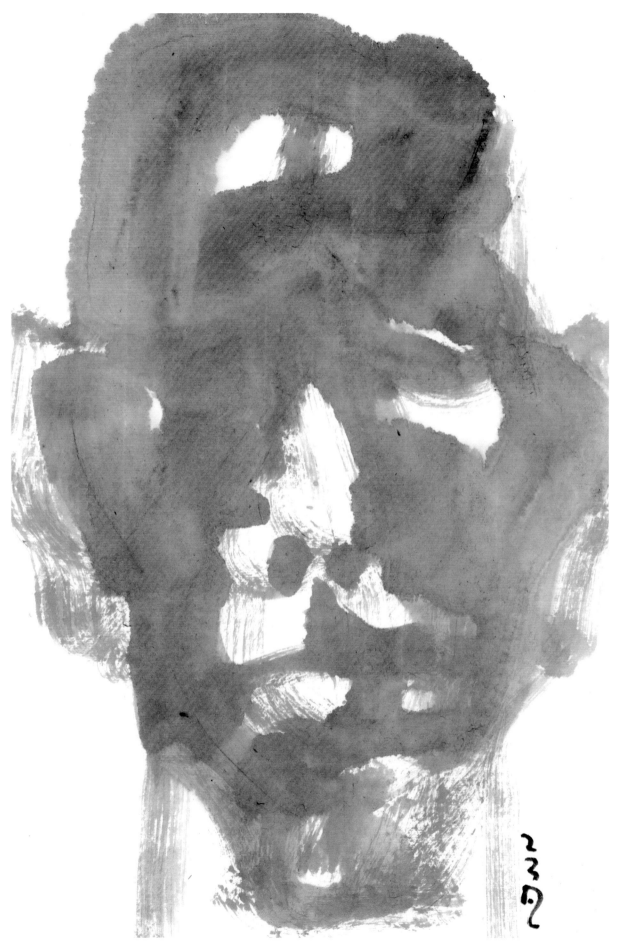

实验水墨
绘画材料：宣纸板，颜料

Experimental Ink Wash Painting
Artwork: Rice paper board, Pigment
33cm × 46cm
2023

实验水墨
绘画材料：宣纸板，颜料

Experimental Ink Wash Painting
Artwork:Rice paper board, Pigment
33cm×46cm
2023

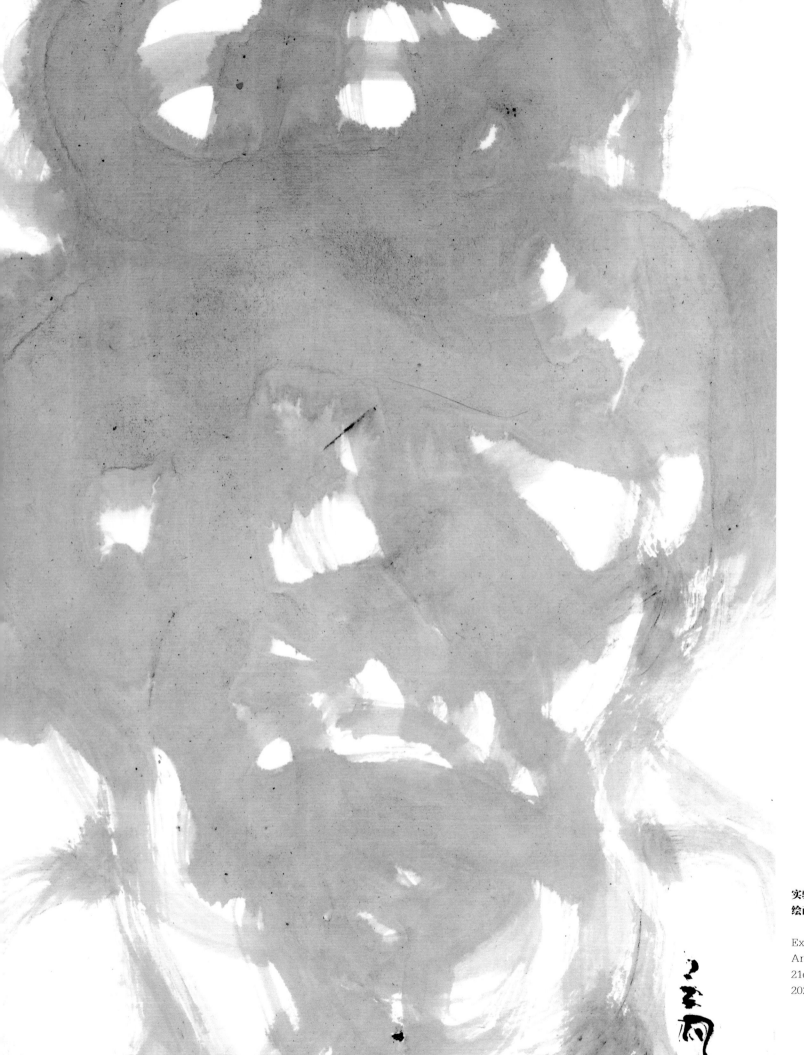

实验水墨
绘画材料：素描纸，水墨

Experimental Ink Wash Painting
Artwork:Sketch paper, Ink
21cm×28cm
2023

实验水墨
绘画材料:素描纸,水墨
Experimental Ink Wash Painting
Artwork:Sketch paper, Ink
21cm × 28cm
2023

实验水墨
绘画材料：复印纸，水墨、签字笔

Experimental Ink Wash Painting
Artwork:Duplicating paper, Ink, Roller pen
21cm×28cm
2023

实验水墨
绘画材料：复印纸，水墨
Experimental Ink Wash Painting
Artwork:Duplicating paper, Ink
28cm×21cm
2023

实验水墨
绘画材料：特种纸，水墨

Experimental Ink Wash Painting
Artwork:Specialty paper, Ink
28cm × 21cm
2023

实验水墨
绘画材料：特种纸，水墨
Experimental Ink Wash Painting
Artwork:Specialty paper, Ink
21cm × 28cm
2023

"人工智能是无法理解这种抽象的情感的,

它只能复刻具体的、既定的形象,

但情感是流动的、变化的,也正是因此,

人类艺术家体察生活本质的能力在人工智能时代变得尤为重要。"

"However, artificial intelligence cannot understand these abstract emotions.

It can only replicate concrete, established images,

but emotions are fluid and changing.

Therefore, the ability of human artists to perceive the essence of life

becomes particularly important in the era of artificial intelligence."

实验水墨
绘画材料：素描纸，水墨

Experimental Ink Wash Painting
Artwork: Sketch paper, Ink
21cm×28cm
2023

实验水墨
绘画材料：复印纸，水墨

Experimental Ink Wash Painting
Artwork: Duplicating paper, Ink
21cm × 28cm
2023

实验水墨
绘画材料：特种纸，水墨

Experimental Ink Wash Painting
Artwork:Specialty paper, Ink
21cm×28cm
2023

实验水墨
绘画材料：宣纸，水墨
Experimental Ink Wash Painting
Artwork:Rice paper, Ink
21cm×28cm
2023

实验水墨
绘画材料：餐巾纸，水墨，水彩

Experimental Ink Wash Painting
Artwork:Napkin paper, Ink, Watercolor
12cm × 14cm
2023

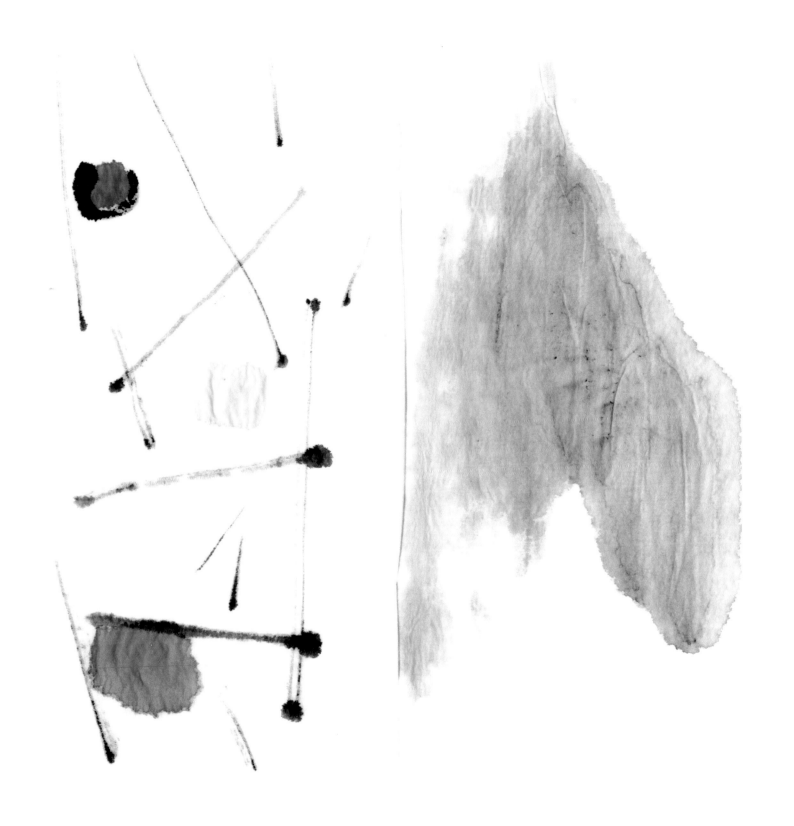

实验水墨
绘画材料：餐巾纸，水墨，水彩

Experimental Ink Wash Painting
Artwork: Napkin paper, Ink, Watercolor
12cm × 14cm
2023

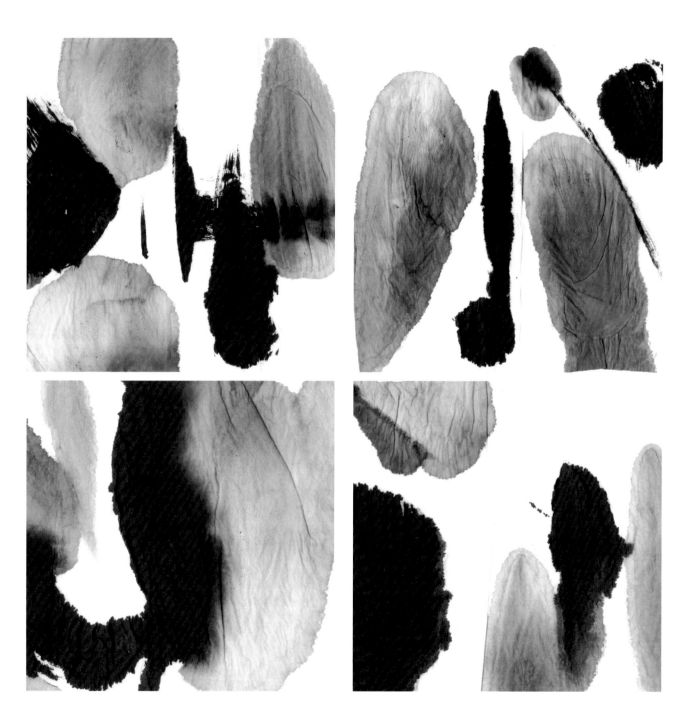

实验水墨　实验水墨
绘画材料：餐巾纸，水墨　绘画材料：餐巾纸，水墨

Experimental Ink Wash Painting
Artwork: Napkin paper, Ink
12cm×14cm
2023

Experimental Ink Wash Painting
Artwork: Napkin paper, Ink
12cm×14cm
2023

实验水墨　实验水墨
绘画材料：餐巾纸，水墨　绘画材料：餐巾纸，水墨

Experimental Ink Wash Painting
Artwork: Napkin paper, Ink
12cm×14cm
2023

Experimental Ink Wash Painting
Artwork: Napkin paper, Ink
12cm×14cm
2023

写生习作
绘画材料：信封纸，签字笔

Life Drawing Practice Painting
Artwork：Envelope paper, Roller pen
10cm × 16cm
2023

实验水墨
绘画材料：素描纸，水墨

Experimental Ink Wash Painting
Artwork: Sketch paper, Ink
33cm × 46cm
2022

实验水墨
绘画材料：复印纸，水彩
Experimental Ink Wash Painting
Artwork: Duplicating paper, Watercolor
21cm×28cm
2021

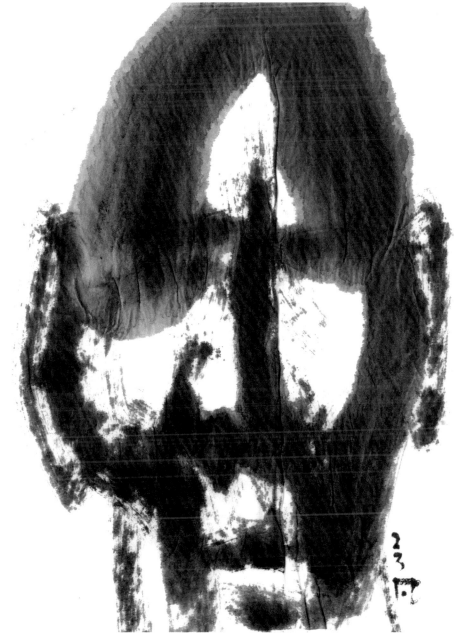

实验水墨
绘画材料：餐巾纸，水墨

Experimental Ink Wash Painting
Artwork: Napkin paper, Ink
12cm × 14cm
2023

实验水墨
绘画材料：餐巾纸，水墨

Experimental Ink Wash Painting
Artwork: Napkin paper, Ink
12cm × 14cm
2023

实验水墨
绘画材料：餐巾纸，水墨

Experimental Ink Wash Painting
Artwork: Napkin paper, Ink
12cm × 14cm
2023

实验水墨
绘画材料：宣纸板，水墨

Experimental Ink Wash Painting
Artwork: Rice paper board, Ink
33cm × 46cm
2023

实验水墨
绘画材料：宣纸板，水墨

Experimental Ink Wash Painting
Artwork: Rice paper board, Ink
33cm × 46cm
2023

实验水墨
绘画材料：速写纸，水墨

Experimental Ink Wash Painting
Artwork: Sketching paper, Ink
21cm×28cm
2023

"与其悲叹'技术吞噬人类文明',

不如思考人类无法被技术取代的优势是什么,

以及人类的优势如何与人工智能实现优势互补。"

"Instead of lamenting that 'technology is devouring human civilization',

we should consider what advantages humans have that cannot be replaced by technology

and how human advantages can complement those of artificial intelligence."

实验水墨
绘画材料：速写纸，水墨

Experimental Ink Wash Painting
Artwork:Sketching paper, Ink
28cm×21cm
2023

写生习作
绘画材料：速写纸，炭笔
Life Drawing Practice Painting
Artwork: Sketching paper, Charcoal
28cm × 21cm
2023

实验水墨
绘画材料：复印纸，水墨

Experimental Ink Wash Painting
Artwork: Duplicating paper, Ink
28cm × 21cm
2023

实验水墨
绘画材料：素描纸，水墨、炭笔

Experimental Ink Wash Painting
Artwork:Sketch paper, Ink, Charcoal
21cm×20cm
2023

实验水墨
绘画材料：牛皮纸，水墨

Experimental Ink Wash Painting
Artwork: Kraft paper, Ink
55cm × 60cm
2023

实验水墨
绘画材料：牛皮纸，水墨
Experimental Ink Wash Painting
Artwork：Kraft paper, Ink
55cm×60cm
2023

构建人工智能时代的东方审美

文 / 孙立军

作为一种面向未来的主体构想与当下的技术实践，人工智能已经引起人类社会政治、经济与文化的技术迭代、形态升级与思维裂变。仅从2023年引起热议的自然语言处理模型ChatGPT和生成式图形图像处理程序Midjourney来看，这一类人工智能基于数据采集、数据分析、模型建立、数据生成和输出控制等程序已经能够创造出与人类创造力相差无几的文本与图像。但值得注意的是，这些人工智能的模型建立首先取决于大数据采集，因此生成式人工智能前期数据的标准、容量、构成与意识形态导向决定着后期的内容生成。有资料显示："ChatGPT的中文语料数据占比为0.09905%，而英文为92.64708%，训练数据不足导致中文答案准确度偏低。"[1]但这一论断还局限于"答案准确度"层面，未能涉及生成内容的中西价值观及其带来的民族文化安全，甚至是人类文明的主体性等深层次问题。因为大部分人工智能是通过开源数据库来完善数据系统并强化学习能力的，而过载的西方文化数据模型及其自身迭代产生的数字文明就意味着"当ChatGPT以自身文化立场对用户的提问作出回答时，可能造成互联网大范围的知识体系及意识形态攻击隐患。"[2]因此，人工智能内容生成隐含着一种西方文明对东方文明、数字文明对自然文明的文化延异危机。由此，本书以AI水墨绘画实践为基准，从东方文化多样性甚至人类自然文明的角度提出东方审美的重大问题。

一、生成式人工智能所隐含的文化延异危机

首先，基于西方文化背景的生成式人工智能对非英语的语言文字体系带来了解构与模糊的文化风险。语言是构成民族文化要素、促进群体内部交流、凝聚民族共识、疏通民族情感的重要媒介，而文字是人类文明的基本符号与显著标志，肩负着传承民族文化精神的重要作用。以汉字为例，"汉字是世界上产生较早的文字。是中华民族最伟大的发明之一，是传承和弘扬中华文化的重要载体，是中华民族的基本标识，也是中华文明的显著标志。"[3]因此，汉字对中国的国家统一、民族共识及文化传播具有高度的战略意义。但是在Midjourney等生成式人工智能的操作实践中，因为训练数据缺乏足够的中文字体样本，因此AI系统在生成包含中文字体的图像时容易导致生成的中文字体出现编码错误或无法正确识别，从而不仅影响图像的功能，还可能引发文化误解。此外，Midjourney的口令数据库亦无法精准解析中

---

1 智通财经《数据要素迈入快车道 AI+传媒引发追捧》，https://www.163.com/dy/article/I22LR5HI05198UNI.html，2023年8月25日访问。
2 李建伟，张书晋《人工智能与国家安全》，《检察风云》2023年第10期。
3 党怀兴《传承汉字文明，坚持文化自信》，《人民论坛》2017年第27期。

文语素，需要将含义丰富的中文语句先大幅压缩，去除分析模型难以理解的形容词与逻辑关系词，将中文语句简化为直白短句，甚至关键词信息，削弱中文语言的文化深度与文化表现力。同时，被压缩的中文语句在程序输入时，还需要将其自动翻译为相对应的英文词汇，这不仅导致中文语素的内涵流失，还会导致中文语言的情感表达最终被英文的语言媒介所代替，并生成出基于英文数据库的"伪中国式内容模型"。这一"伪中国式内容模型"也将随着 Midjourney 等西方技术的传播而在全球普及，最终挤压中国正统文化的生存空间，从而驱使全球多样文明驶入以英语为中心的单极化文明体系。

同时，基于西方文化背景的人工智能技术还对其他文化的经典视觉符号带来混杂与错用的风险，消解其原有的文化意义。以"龙"为例，"'龙'作为中华民族的文化标志与精神图腾，在封建时代，它象征着不可触逆的神圣皇权，在当代中国，它代表着尊贵与祥瑞。"[4] 同时，龙的形象经过长期历史演进，已经形成相对稳定的造型结构。明代医学家李时珍在《本草纲目》中提道："龙有九似，头似驼，角似鹿，眼似鬼（鬼王相），耳鼻似牛，颈似蛇，腹似蜃（蟒），鳞似鱼，爪似鹰，掌似虎。"[5] 因此，在中国的传统造型设计中，龙的视觉元素基本遵循以上组成原则。但西方的"龙"在形象与意义方面却和中国龙存在较大差异：其"通常都有翅膀，骨头是中空的，但很强壮，质量轻。有非常复杂的肌肉系统。除了脖子和腹部，龙的全身都覆盖着发光的鳞片。它用四只强有力的脚步行，用一对像蝙蝠翼的巨翼飞行。"[6] 且西方的龙常常与警戒相联系，"是人类的敌人……无论是传说中的 dragon，还是 dragon 这个单词本身中所具备的一些含义，都具有凶狠甚至邪恶的色彩"。[7] 因此，基于西方文化背景的 Midjourney 生成的"龙"的造型更多地偏向西方对龙的视觉想象，从而在涉及中国龙的生成过程中，"西方龙"替换掉了"中国龙"的原有形象和意义。

实际上，人工智能所带来的文化风险不仅是对非西方文明的文化多样性问题，而是对包含西方文明在内的整个人类自然文明。西方学者已经"发现在训练中使用模型生成的内容会导致生成的模型出现不可逆转的缺陷，原始内容分布的末端会消失……最终导致模型崩溃。"[8] 即，人工智能创造的数字文明可能会反噬人类的自然文明，从而导致自然文明的消

4　王村杏，孙立新《"中国龙"构成设计的造型分析》，《艺术评论》2012年第10期。
5　李时珍《本草纲目》，北京：北京燕山出版社2007年版，第34页。
6　赵丽玲，王杨琴《论中西方龙形象及其文化意义的差异》，《湖北工业大学学报》2010年第6期。
7　肖达娜《中国龙形象在西方文化中的重构》，《南方论刊》，2015年第8期。
8　Ilia Shumailov. Et, The Curse of Recursion: Training on Generated Data Makes Models Forget. arXiv:2305.17493（31 May 2023），pp.1-18.

解与变异。从抽象思维到语言，再到文字，人类的思想传播已经经过二度文化塑形，即自然语言并不能充分表达人类的思维与情感，而文字又无法完整地表达语言，因此语言和文字本身就存在着传播折扣。人工智能在构建"人机传播"的机器语言过程中，又在人人传播的自然语言基础上增加了一道新的传播折扣，这一传播折扣反过来也将影响原有的自然语言系统，从而异化人类自然文明。因此，尽管人类自然文明的消解和变异首当其冲的是非西方文明，但是最终，单极化的西方文明本身也会因为数字文明的几何迭代而被绑上"加速主义"的失控列车，从而驶向不可知、不可测的"人类世"终局。

二、以东方审美的文化底蕴丰富人类文明的生命力与多样性

人工智能是现代信息文明的技术产物，其内容生产不仅不可避免地裹挟着西方意识形态，更暗含着异化整个人类自然文明的文化危机。审美是文化形态的一部分，如何以审美激活数字文明背景下人类自然文明的生命力，并以此增加人类自然文明的多样性，还需要从东方文化中寻找美学智慧。其中，东方审美是基于中华文明而产生的艺术美学价值体系。"中华民族在长期的生存与发展过程中形成了自己的美学精神，这种精神既是中华民族生存与发展的结晶，又是中华民族生存与发展的原动力。"[9]因此，中华文明促成了独具特色的东方审美的发生，并深刻影响整个东亚地区的民族民间文学、音乐、绘画、雕塑、舞蹈，甚至电影、动画与新媒体等各个艺术门类的观念形成。例如，中国古典哲学中的"无为之为""中和之美""天人合一""雅俗之辩"等美学思想，催生了东方传统艺术中"气韵生动""应物象形""臻于化境""吟风颂雅"等美学精神。因此，对于东方审美，不仅要以现代性的技术视野，更应该以中华文明作为文化基底才能深刻把握其内涵特征。"中华文明具有突出的连续性、突出的创新性、突出的统一性、突出的包容性、突出的和平性。"[10]因此，基于中华文明的突出特性，东方审美具有连续性、创新性、统一性、包容性与和平性五大底蕴。

首先，东方审美是具有连续性的美学体系。在长达五千多年的中华美学发展过程中，以艺术为载体的中华文明从未中断，中华精神始终在各个艺术门类中赓续发展。水墨绘画是独具东方美学特色的艺术形式，早在战国时期便出现了水墨画的雏形，彼时主要以线描为主，辅以设色的人物画像。到了魏晋南北朝，随着美学思想和绘画理论的成熟，水墨画开始有了系统发展。顾恺之、谢赫等人的绘画理论著作，如谢赫的《古画品录》中的"六法"，为水墨画的创作提供了理论指导。本画册在当代艺术多元发展的时代背景下，以水墨为媒介对中国古典意境美学进行再探索，彰显出中国文人一脉相承的精神品位。这种对历史的连续性的承接赋予了本书东方审美以浓郁的东方文化风格，从而贯通古代中国、现代中国和未来中国绵延不绝的美学精神。

其次，东方审美是具有创新性的美学体系。面对不断变化的时代趋势，东方美学以创新作为最大的文化路径不断发展自身，从而体现出强烈的与时俱进的精神和革故鼎新的勇气。除《牧笛》《山水情》等传统水墨动画片之外，近年来的《立秋》《秋实》等数字水墨动画片均积极而审慎地拥抱当代美学，以数字技术升级推进水墨美学创新，笔法空灵，意境优美，通过8K超高清视频技术复现并创新齐白石水墨绘画的艺术韵味，阐释水墨绘画、数字技术与影像艺术融合发展的时代缩影。艺术家在绘

9　陈望衡《中国美学精神简论》，《中州学刊》，2021年第6期。
10　习近平《在文化传承发展座谈会上的讲话》，http://www.qstheory.cn/dukan/qs/2023-08/31/c_1129834700.htm，2023年8月28日访问。

制本画册老人头像的过程中，摒弃了传统的写意和工笔技法，转而采用一种更有力道的表现手法。这种手法既结合了西方的素描技巧、现代艺术的构成原理，以及中国传统书法的笔力运用，创造出一种新的水墨语言。这种力道的表达挑战传统水墨柔和、含蓄的风格特点，展现了一种更加直接和强烈的情感表达。因此，这些水墨作品不仅传承中华文明的本土精神，更用不断创新续写着东方审美的文化传奇。

再次，东方审美是具有统一性的美学体系。中国"是统一的多民族国家，中华民族多元一体是我国的一个显著特征。多元意味着丰富的文化形式，一体则展现为中华文化独一无二的理念、智慧、气度、神韵"。[11]因此，东方审美的统一性体现在中国艺术既有丰富多彩的体裁样式，又有明确统一的家国追求，在面对他文化的殖民实践中能够凝练出高度的主体意识和自觉的抗争精神。在20世纪40年代的国家危难之际，中国水墨艺术家们以东方审美作为精神旨归，以"新国画"作为媒介武器，通过《三灶岛外所见》《渔民之劫》《从城市撤退》《中山难民》《侵略者的下场》等抗战题材水墨绘画，"融汇时代精神、民族风格，突出画家深沉的爱国情感"[12]和对中华文明内部统一性的高度认可。因此，东方审美的统一性既给予了中国各民族、各文化之间交流相融的精神纽带，也在面对异域、异端文明时为中华文明的发展提供了凝聚精神的文化指引，更为数字文明背景下的"人类命运共同体"铺垫了理论根基。

又次，东方审美是具有包容性的美学体系。中华大地广袤复杂的地理空间孕育了丰富多彩的文化类型，而东方美学的形成就是中华文明不断兼容并蓄不同文化成果的历史过程。尽管水墨绘画是中国独特的艺术形式，但它在传统的基础上，不断吸收外来艺术元素，如西方的色彩理论、构图技巧等，使得水墨画在保持传统韵味的同时，也展现出现代艺术的活力，这种开放性和创新精神，使得水墨画在国际艺术舞台上具有更广泛的吸引力。近年来，凡悲鲁的"宇宙"系列作品便是这一探索的杰出代表。这些作品不仅吸收康定斯基构成主义的精髓，通过抽象的形式和色彩的大胆运用，打破传统水墨画的界限，创造出一种全新的视觉语言。在"宇宙"系列中，凡悲鲁巧妙地将水墨的流动性与西方抽象艺术的几何构成相结合，使得画面既有东方的诗意与哲思，又蕴含着西方现代艺术的理性与秩序。这种风格的形成，不仅丰富水墨画的表现力，也为世界艺术界提供一种独特的东方视角，展现中国艺术家在全球化背景下的文化自信和创新能力。因此，包容性决定了东方审美和人类自然文明在人工智能时代的赓续具有源源不断的生命力。

最后，东方审美是具有和平性的美学体系。"中华文化崇尚和谐，中国'和'文化源远流长，蕴含着'天人合一'的宇宙观、'协和万邦'的国际观、'和而不同'的社会观、'人心和善'的道德观。"[13]这样的"和"思想始终贯穿在中华文化艺术的价值体系之中，并以此形成东方审美的关键特征。北宋画家王希孟的《千里江山图》描绘了壮丽的江山景色，展现了人与自然和谐共存的理想境界。画中的山水、建筑、人物和船只等元素，构成了一幅生机勃勃的自然画卷，体现了人与自然和谐相处的美好愿景。而元代画家赵孟頫的《鹊华秋色图》则描绘了济南附近的鹊山和华不注山，以及山脚下的村舍、树丛和浅滩。画面中的自然景观与人文景观交织，体现了人与自然和谐共存的主题。而在本书中，艺术家利用AI技术，模拟出水墨画的笔触、墨色变化和构图，甚至

11　白阳，徐壮《从中华文明统一性看中华民族的凝聚力》，https://www.spp.gov.cn/tt/202306/t20230614_617916.shtml，2023年8月25日访问。
12　马刚 等《抗战时期国画家在重庆地区的艺术活动及相关美术研究》，《内江师范学院学报》2023年第1期。
13　人民论坛"特别策划"组《中华历史文化所孕育的和平基因》，《人民论坛》2017年第21期。

创造出超越艺术家手绘水墨画表现力的新视觉效果。在这种"人机合一"的美学状态下，艺术家的手绘作品与 AI 创作的作品相互补充，共同构成了一个多元化的艺术世界。手绘作品保留了艺术家的情感和直觉，而 AI 作品则展现了算法的逻辑和效率。两者的结合，不仅挑战了传统艺术创作的定义，也为水墨画创作"人机合一"的未来发展提供了新的方向。

三、构建人工智能时代东方审美的文化路径

当下的人工智能技术仍在加速迭代，人类自然文明已不可避免地卷入了数字文明所带来的存亡危机。"一个新的文化时代的到来总会引起人们的恐慌、疑虑、抵制、不适，这是必然现象，重要的是我们没有回头路，必须向前走。"[14] 从长远来看，人工智能技术必然会迫使人类文明驶向单极化的不归之路，从而最终以数字文明挤压和取代自然文明。正如霍金判断："人类需警惕人工智能发展威胁。因为人工智能一旦脱离束缚，以不断加速的状态重新设计自身，人类由于受到漫长的生物进化的限制，将无法与之竞争，从而被取代。"[15] 因此，基于中华民族现代文明提出构建东方审美的宏观意义，不仅是要提醒人们注意西方技术隐含的中西二元对立的意识形态风险，还要最大限度地以人类最淳朴的审美感性对抗数字文明的机械理性，以"人类命运共同体"的宏观视野对抗人类文明单极化的趋势，以对艺术精神的终极追求达成对物质世界的脱离与皈依。因此，要以超越的哲学姿态和虔诚的生命美学构建人工智能时代的东方审美，这也正是本书深度应用人工智能进行水墨绘画创作的意义之一。

要构建人工智能时代的东方审美，必须首先坚定中华文明的文化自信。五千年来，"中华优秀传统文化中蕴含的天下为公、民为邦本、为政以德、革故鼎新、任人唯贤、天人合一、自强不息、厚德载物、讲信修睦、亲仁善邻等思想，既是人们在长期生产生活中积累的宇宙观、天下观、社会观、道德观的重要体现，也是中华优秀传统文化的精华资源"。[16] 这些文化资源不仅支撑着中华民族在物质层面生生不息，繁衍至今，而且在多次民族存亡的重大危机中给予人民以无限的希望和勇气。在本书中，艺术家塑造了大量的农民形象，而在中国传统文化中，农民形象被赋予了吃苦耐劳、勤奋朴实的品质，无论是农民耕作、工匠制作还是士人治学，都强调勤奋努力。《诗经》中有"夙兴夜寐，毋忝尔所生"的诗句，表达对勤劳的赞美。在中国数千年的农耕文明中，农民作为社会的基础，承担着养育国家、维护社会稳定的重要角色。因此，本书中的农民是中华民族繁衍生息的文化底色与文化标志。而从农民形象中所传达出来的"独一无二的理念、智慧、气度、神韵，增添了中国人民和中华民族内心深处的自信和自豪"。[17] 因此，要以中华文明的文化自信坚定对东方审美的艺术自信。

要构建人工智能时代的东方审美，还要秉持开放包容的艺术态度。"包容性"既是中华文明和东方审美的突出特性，也是在人工智能技术不断升级和数字文明逐渐成形的背景下，促进东方审美不断丰富的重要途径。在数字文明诞生之前的中西方文明交流中，无数的哲学家、思想家便深刻地论述了开放包容对一种文

14　庞井君《以审美方式探寻人类精神体系之根》，《人民论坛·学术前沿》2017 年第 10 期。
15　王心馨《霍金北京演讲：人工智能也可能是人类文明的终结者》，https://www.thepaper.cn/newsDetail_forward_1672326，2023 年 8 月 25 日访问。
16　宇文利《新时代推进文化自信自强》，《新视野》2023 年第 3 期。
17　习近平《在中国文联十大、中国作协九大开幕式上的讲话》，北京：人民出版社 2016 年版，第 4 页。

明、文化和审美的重要性。梁启超认为:"拿西洋的文明来扩充我的文明,又拿我的文明去补助西洋的文明,叫他们化合起来成一种新文明。"[18] 英国哲学家伯特兰·罗素也认为,中西文化包容对文明的长远发展具有积极意义:"他们可以从我们这里学到必不可少的实用的效率;而我们则可以从他们那里学到一些深思熟虑的智慧。"[19] 因此,开放包容是"人类文明进步的内生动力,没有任何一种文明可以在完全摆脱异质文明的前提下孤立发展"。[20] 在当前以人工智能技术为基底的数字文明时代,大数据管理、算法艺术、奇点艺术、智能交互艺术、纳米艺术、智能打印艺术、自动化生成、模拟仿真、人机协同甚至"人机一体"等创作方式与数字风格都会逐渐成熟并普及,而东方审美的包容性注定了其会以积极的姿态拥抱不同的文化特征和创作方式,内融于自身的文脉之中,从而以东方文明真正激发人类自然文明的生命力。

要构建人工智能时代的东方审美,还必须不断推进中华文明自身的守正创新。中华文明是东方审美的文化基底,中华文明能够守正创新,就意味着东方审美具有无限发展的文化动力。其中,"守正"就是要在艺术领域"守马克思主义文艺理论的导向功能之'正',要守中华优秀传统文化创造性转化、创新性发展之'正',守中华美学风尚、中华美学精神之'正'"[21],守中国艺术的民族立场和求真、求美、求善之"正";创新则是要在美学理念、创作方法、评价标准、理论水平等方面与时俱进,践行"苟日新、日日新、又日新"的创新精神。在本书中,传统的水墨画创作依赖艺术家的手工技艺和个人体验,而生成式人工智能的介入,使得艺术创作过程更加多样化。艺术家通过设定算法参数,让 AI 学习并模拟水墨画的风格,从而创造出具有独特风格的作品。这种过程改变了艺术家与作品之间的关系,以及观众对艺术创作过程的理解,从而实现了水墨画创作过程的守正创新。

结语

人工智能时代已经到来,它不仅给艺术领域带来了一系列生产问题,更重要的是其隐含了人类文明整体性的文化危机。目前,国内已经注意到了生成式人工智能或将导致的民族文化安全问题,有政协委员呼吁:"支持规范国产化 ChatGPT 研发及应用。"[22] 本书亦意图站在人类文明的哲学高度思考如何以水墨绘画的东方审美平衡中西方文明的多元化,并以此促进人工智能时代人类自然文明的赓续问题。其中,构建内涵丰富、特色鲜明的东方审美是重要的文化路径。要构建人工智能时代的东方审美,首先需要以中华文明作为根基,找准东方审美的连续性、创新性、统一性、包容性与和平性五大文化底蕴;其次,要以文化自信、开放包容和守正创新作为东方审美的具体路径,以此通达中西的意识形态对立及人类自然文明的延续问题。在人工智能技术即将在艺术创作领域应用普及的时代前夜,我们不仅要积极面对、深入研究人工智能,更应该勇于担当时代使命,用优秀的文艺作品彰显东方审美旨趣,以深度应用了当代人工智能技术的艺术形式弘扬东方美学精神,以此推动人类自然文明与数字文明更好的平衡。

---

18 梁启超《饮冰室合集·专集第 23 册》,北京:中华书局 1989 年版,第 55 页。
19 [英]伯特兰·罗素《中国问题》,秦悦译,上海:学林出版社 1996 年版,第 146-156 页。
20 刘志刚《从"文明冲突论"到人类命运共同体——中西方对待文明冲突的不同逻辑》,《学术界》2021 年版,第 207 页。
21 段吉方《论中国特色文艺评论的"守正创新":理论、路径与任务》,《社会科学辑刊》2022 年第 4 期。
22 高志民《9 位全国政协委员共同呼吁:支持规范国产化 ChatGPT 研发及应用》,http://www.rmzxb.com.cn/c/2023-05-24/3350234.shtml,2023 年 8 月 29 日访问。

Construct the Oriental Aesthetic in the Era of Artificial Intelligence

Sun Lijun

As a future-oriented subjective concept and current technological practice, AI has already aroused technological iteration, morphological upgrading and thinking fission in socio-political, economic and cultural contexts. Judging from the natural language processing model ChatGPT and the generative graphic image processing programme Midjourney, both of which had already been hotly debated in 2023, this type of AI is already capable of creating texts and images that are comparable to human creativity based on procedures such as data collection, data analysis, model building, data generation and output control. However, it is worth noting that the model building firstly depends on big data collection, so the standard, capacity, composition and ideological orientation of the data in the early stages of generative AI determines the content being generated in later stages. It has been shown that "ChatGPT's Chinese corpus data accounted for 0.09905%, while the English was 92.64708%, which directly led to the low accuracy of corpus-based answers generated in Chinese." [1] This assertion remains confined to the dimension of "answer accuracy" and does not extend to issues concerning the synthesis of Eastern and Western values within the generated content, nor does it address deeper concerns such as the safeguarding of national cultural integrity and the agency of human civilization. Because the majority of artificial intelligence refine their data systems and augment their learning capabilities through open-source databases, the predominance of Western cultural data models and the digital civilization iteratively generated from them signifies that "when ChatGPT responds to users' inquiries from its own cultural stance, it may pose a risk of widespread negative influence on the knowledge systems and ideological constructs of the internet." [2] Thus, it can be said that AI-generated content implies a cultural crisis of Western civilization over Eastern civilization, and of digital civilization over natural civilization. Consequently, this book, based on the practice of AI ink painting, raises

---

1   WisdomTree, "Data elements move into the fast lane AI+media sparks pursuit", https://www.163.com/dy/article/I22LR5HI05198UNI.html, accessed 25 August 2023
2   Li Jianwei, Zhang Shujin "Artificial Intelligence and National Security", Procuratorate Wind and Cloud 2023, No. 10.

significant issues concerning Eastern aesthetics from the perspectives of Eastern cultural diversity and even human civilization.

I . The Implicit Cultural Crisis within Generative Artificial Intelligence

Firstly, generative artificial intelligence, based on Western cultural backgrounds, brings about deconstruction and vagueness in the existing systems of non-English languages. Language serves as a crucial medium for constituting national cultural elements, facilitating intra-group communication, consolidating national consensus, and mediating national emotions, while writing acts as a basic symbol and prominent sign of human civilization, bearing the significant task of transmitting the spirit of national culture. Taking Chinese characters as an example." Chinese characters are the earliest characters in the world. It is one of the greatest inventions of China, an important carrier for the inheritance and promotion of Chinese culture, the fundamental identifier and the distinctive sign of the Chinese nation." [3] Therefore, Chinese characters hold a high strategic significance for China's national unity, national consensus, and cultural dissemination. However, in the operational practices using generative AI like Midjourney, due to a lack of samples of Chinese typefaces in the training data, AI systems are very likely to produce encoding errors or fail to correctly recognize Chinese characters when generating images containing them, thus not only affecting the functionality of the images but can also lead to cultural misunderstandings. Additionally, Midjourney's current command database can't precisely parse Chinese morphemes, requiring that richly meaningful Chinese sentences be significantly compressed, stripping away adjectives and logical conjunctions that proves to be so hard for the analysis model to understand accurately, and simplifying the Chinese statements into blunt short sentences, or even just keywords, diminishing the cultural depth and expressive potential of the Chinese language. Moreover, when these compressed Chinese sentences are input into the program, they also need to be automatically translated into corresponding English terms, leading not only to a loss of the connotations of Chinese morphemes but also a replacement concerning the emotional expressions, thus generating a "pseudo-Chinese content model" based on what exactly is an English database, which will also spread globally with the dissemination of Western technologies like Midjourney, ultimately squeezing the living space of orthodox Chinese culture and propelling the world's diverse civilizations into a unipolar civilization system centered around English.

Simultaneously, artificial intelligence technologies based on Western cultural backgrounds also pose risks of confusion

---

3 Dang Huaixing, "Inheriting the civilisation of Chinese characters and upholding cultural confidence", People's Forum, No. 27, 2017.

and misuse concerning the classic visual symbols of other cultures, diluting their original cultural significance. Take the "dragon" as an example, "'dragon' as the cultural symbol and spiritual totem of the Chinese nation, in the feudal era, it symbolized the untouchable, sacred imperial power, and in contemporary China, it represents nobility and auspiciousness." [4] At the same time, the image of the dragon, through its long historical evolution, has formed a relatively stable composition structure. Ming Dynasty medical expert Li Shizhen mentioned in the Pen-ts'ao Kan-mu: Compendium of Materia Medica: "The dragon has nine likenesses, head like a camel, horns like a deer, eyes like a ghost (Ghost King), ears and nose like a cow, neck like a snake, belly like a mirage (python), scales like a fish, claws like an eagle, palm like a tiger." [5] Therefore, in traditional Chinese character design, the visual elements of the dragon generally adhere to the above structural principles. However, the Western "dragon" significantly differs in both the looking and the meaning from the Chinese dragon: "It usually has wings, hollow yet strong bones, and a light body with a complex muscular system. Except for the neck and abdomen, the dragon's body is covered in glowing squama. It walks on four strong legs and flies with a pair of wings like those of a bat." [6] And dragons in the West are often associated with vigilance, "they are seen as a foe of mankind, embodying fierceness and even evil traits in both the mythological 'dragon' and the connotations of the word itself." [7] Thus, when generative models like Midjourney, grounded in Western cultural backgrounds, create images of "dragons," they tend to reflect Western visual interpretations of the dragon, which in the process of generating images of the Chinese dragon, replaces the original image and meaning of the "Chinese dragon" with that of the "Western dragon."

In fact, the cultural risks brought about by artificial intelligence extend not only to the issue of cultural diversity within non-Western civilizations but also encompass the entirety of human natural civilization, including Western civilization. Western scholars have found "The use of model-generated content in training causes irreversible defects in the resulting models, where tails of the original content distribution disappear. We refer to this effect as model collapse." [8] This suggests that the digital civilization created by artificial intelligence might devour human natural civilization, thus causing the dissolution and mutation of natural civilization. From abstract thinking to language pattern, and then to characters, the spiritual dissemination of human has

---

4　Wang Cunxiong and Sun Lixin, "Stylistic Analysis of the Compositional Design of the "Chinese Dragon"", Art Review, No. 10, 2012.

5　Li Shizhen, Compendium of Materia Medica, Beijing: Beijing Yanshan Publishing House, 2007 edition, p. 34.

6　Zhao Liling, Wang Yangqin "On the differences between the images of dragons and their cultural meanings in China and the West", Journal of Hubei University of Technology, No. 6, 2010.

7　Dana Xiao, "The Reconstruction of the Chinese Dragon Image in Western Culture," Southern Discourse, No. 8, 2015.

8　Ilia Shumailov. Et, The Curse of Recursion: Training on Generated Data Makes Models Forget. arXiv:2305.17493 (31 May 2023), pp. 1-18.

already undergone a cultural re-molding; that is, natural language cannot fully express human spirits and emotions, and writing cannot fully express essence of verbal language, hence there is an inherent transmission discount in language and writing itself. Artificial intelligence, in constructing the machine language for "human-machine communication," adds a new layer of transmission discount atop the natural language used in human-to-human communication. This transmission discount, in turn, will affect the existing natural language system, thus alienating human natural civilization. Therefore, although the dissolution and mutation of human natural civilization initially impact non-Western civilizations the most, eventually, even the unipolar Western civilization itself will be bound to the runaway train of "accelerationism" due to the geometric iterations of digital civilization, thus steering towards an unknowable, unpredictable "anthropocene" finale.

II. Enriching the Vitality and Diversity of Human Civilization with the heritage of Eastern aesthetics

Artificial intelligence, a technological product of modern information civilization, not only inevitably involves Western ideologies but also subtly implies a cultural crisis that alienates the entirety of ecological civilization. Aesthetics is a part of cultural forms, and how to invigorate the vitality of ecological civilization under the backdrop of digital civilization, and thereby enhance its diversity, requires seeking aesthetic wisdom from Eastern culture. In this context, we are going to focus on Eastern aesthetics that are derived from the artistic and aesthetic values produced within Chinese civilization. "The Chinese nation has formed its own aesthetic spirit in the course of its long-term survival and development, which is both the crystallisation of the survival and development of the Chinese nation and the driving force behind it." [9] Consequently, Chinese civilization has fostered the emergence of distinctive Eastern aesthetics, profoundly influencing the artistic ideals in various artistic categories across the entire East Asian region, including folk literature, music, painting, sculpture, dance, and even film, animation, and new media. For example, aesthetic ideals in Chinese classical philosophy such as "action through inaction," "the beauty of moderation," "the unity of heaven and mankind," and "the distinction between elegance and vulgarity" have inspired the aesthetic spirits of "vivid charm," "pictographic," "attaining aura," and "chanting the poetry, odes and hymns" in traditional Eastern arts. Therefore, understanding Eastern aesthetics requires not only a modern technological perspective but also a precise understanding of its inherent features based on Chinese civilization as the cultural foundation. "Chinese civilization is characterized by

---

9 Chen Wangheng, "A Brief Discussion of the Spirit of Chinese Aesthetics," Zhongzhou Xuejian, No. 6, 2021.

remarkable continuity, innovation, unity, inclusiveness, and peacefulness." [10] Based on these prominent characteristics of Chinese civilization, Eastern aesthetics embodies five foundational qualities of continuity, innovation, unity, inclusiveness, and peacefulness.

Firstly, Eastern aesthetics is a continuous aesthetic system. Throughout the development of Chinese aesthetics over more than five thousand years, Chinese civilization, with art as its medium, has been continuously evolving across various artistic categories. Ink painting is an art form uniquely characterized by Eastern aesthetics, with its prototype appearing as early as the Warring States period, initially dominated by line drawing supplemented by colored figure portraits. By the Wei-Jin and the Northern and Southern Dynasties, with the maturation of aesthetic value and painting theory, ink painting began to develop systematically. Painting theories by artists like Gu Kaizhi and Xie He, such as the "Six Principles" in Xie He's *Record of the Classification of Old Paintings*, provided theoretical guidance for the creation of ink paintings. In the context of the diverse development of contemporary art, this books aims to re-explore classical Chinese aesthetic concepts through the medium of ink, highlighting the spiritual taste inherited by Chinese literati. This continuity with history imbued this book with a rich Eastern cultural style, thus linking the aesthetic spirit of ancient China, modern China, and future China continuously.

Secondly, Eastern aesthetics is an innovative aesthetic system. In response to constantly changing trends, Eastern aesthetics continually evolves itself by embracing innovation as the priority, thereby exhibiting a spirit of keeping pace with the times and the courage to reform and innovate. Apart from traditional ink animated films like *The Cowherd's Flute* and *Feelings of Mountains and Waters*, recent digital ink animated films such as *Harvest* and *The Beginning of Autumn* have actively and cautiously embraced contemporary aesthetics, using digital technology to bring the traditional ink aesthetic to the next level, creating a sensual realm that is ethereal, utilizing 8K ultra-high-definition video technology to reproduce and innovate on the artistic charm of Qi Baishi's ink paintings, encapsulating the integration of ink painting, digital technology, and film art as a microcosm of the era. In the process of creating portraits of elderly individuals in this book, the painter had abandoned traditional freehand and meticulous painting techniques and turned to a more forceful expressive approach. This method combines Western sketching skills, modern art's principles of composition, and the use of brushwork in traditional Chinese calligraphy, creating a new ink language. This forceful expression challenges the mild and implicit style of traditional ink painting, showcasing a more direct and intense emotional expression. Therefore, these ink paintings by human artists not only inherit the indigenous spirit of Chinese civilization but also continually innovate, bringing

---

10 Xi Jinping's Speech at the Symposium on Cultural Heritage Development, http://www.qstheory.cn/dukan/qs/2023-08/31/c_1129834700.htm, accessed 28 August 2023.

forward brand new ways of visualizing Eastern aesthetics.

Thirdly, Eastern aesthetics is a unified aesthetic system. China is "a unified multi-ethnic nation, and the multi-ethnic unity of the Chinese nation is a prominent characteristic of our country. Diversity implies rich cultural forms, while unity is manifested in the unique concepts, wisdom, temperament, and charm of Chinese culture".[11] Therefore, the unity of Eastern aesthetics is reflected in Chinese art, which not only possesses a diverse array of genres and styles but also shows a clear and unified pursuit of the nation. It is capable of distilling a high degree of subject consciousness and the spirit of resistance in the face of cultural colonization. During the national crises of the 1940s, Chinese ink artists adopted Eastern aesthetics as their spiritual principle and used "new national painting" as a medium and weapon. Through anti-war themes ink paintings such as *Views Outside Sanzao Island*, *The Plight of the Fishermen*, *Retreat from the City*, *Zhongshan Refugees*, and *The Fate of the Invaders*, "blending the spirit of the era and national artistic style, highlighting the painters' deep patriotic emotions"[12] showing a high degree of recognition of the unity within Chinese civilization. Thus, the unity of Eastern aesthetics not only provides a spiritual bond for integration among China's various ethnic groups and cultures but also offers cultural guidance that consolidates the spirit of Chinese civilization in the face of foreign and heterodox civilizations. Furthermore, it lays a theoretical foundation for the "community of shared human destiny" under the backdrop of digital civilization. Furthermore, Eastern aesthetics is an inclusive aesthetic system. The vast and complex geographical space of China has nurtured a diverse array of cultural forms, and the formation of Eastern aesthetics is a historical process in which Chinese civilization continuously integrates and accumulates different cultural achievements. Although ink painting is a unique art form in China, it continually absorbs foreign artistic elements on a traditional basis, such as Western color theory and compositional techniques, allowing ink painting to maintain its traditional charm while also exhibiting the vitality of modern art. This openness and spirit of innovation make ink painting appealing to the mass majority on the international art stage. In recent years, Fanbeilu's *Cosmos* series is a distinguished example of this exploration. These works not only absorb the essence of Kandinsky's Constructivism through the bold use of abstract forms and colors, breaking the existing aesthetic patterns of traditional ink painting, but also create a new visual language. In the *Cosmos* series, Fanbeilu skillfully combines the fluidity of ink with the geometric compositions of Western abstract art, creating images that possess both the poetic and philosophical essence

---

11　Bai Yang, Xu Zhuang "The Cohesion of the Chinese Nation from the Unity of Chinese Civilisation", https://www.spp.gov.cn/tt/202306/t20230614_617916.shtml, accessed 25 August 2023.

12　Ma Gang et al. "Artistic Activities of National Painters in Chongqing Area during the Anti-War Period and Related Fine Arts Research", Journal of Neijiang Teachers College, No. 1, 2023.

of the East and the rationality and order of Western modern art. The formation of this style not only enriches the expressiveness of ink painting but also provides the global art community with a unique Eastern perspective, showcasing the cultural confidence and innovative capacity of Chinese artists in a globalize context. Therefore, inclusiveness ensures that Eastern aesthetics and ecological civilization continue to have a sustained vitality in the era of artificial intelligence. Finally, Eastern aesthetics is a peace-promoting aesthetic system. "Chinese culture venerates harmony, and the 'harmony' culture of China is ancient and profound, encompassing a cosmology of 'unity between heaven and man,' an international outlook of 'harmonious coexistence among all nations,' a social view of 'seeking common ground while preserving differences,' and a moral philosophy that 'men are born to be kind.'" [13] Such notions of "harmony" consistently permeate the value system of Chinese cultural arts, forming a key characteristic of Eastern aesthetics. The Northern Song painter Wang Ximeng's *A Thousand Li of Rivers and Mountains* depicts the magnificent natural scenery, presenting an ideal state of harmonious coexistence between human and nature. Elements in the painting like landscapes, architecture, people, and boats together create a vibrant natural tableau that embodies a vision of harmonious interaction. Yuan dynasty painter Zhao Mengfu's *Autumn Colors on the Que and Hua Mountains* illustrates the Mount Que and Mount Hua Buzhu that are close Jinan city, as well as the cottages, groves, and shoals at the mountain's base. The intertwined natural and cultural landscapes in the painting reflect the theme of harmonious. In this book, artists utilize AI technology to simulate the brushstroke, ink color variations, compositions of ink painting, and even create new visual effects that surpass the expressiveness of hand-drawn ink paintings. In this "human-machine unity" aesthetic state, the hand-drawn works of the artists complement those created by AI, together constituting a diversified artistic world. The hand-drawn works retain the artist's emotions and intuition, while the AI-generated pieces display the logic and efficiency of algorithms. The integration of both not only challenges the traditional definitions of artistic creation but also offers a new direction for the future development of "human-machine unity" in ink painting creation.

Ⅲ. Constructing a Cultural Path for Eastern Aesthetics in the Age of Artificial Intelligence

Currently, artificial intelligence technology continues to accelerate in its iterations, inevitably drawing human civilization into a crisis of existence brought about by the digital civilization. "The arrival of a new cultural era always induces panic, skepticism, resistance, and discomfort; this is inevitable, and most importantly, there is no turning back—we must move forward." [14] In the long run, AI technology will inevitably

---

13 People's Forum "Special Planning" Group, "The Peace Gene Bred by Chinese History and Culture", People's Forum, No. 21, 2017.

drive human civilization towards an irreversible path of unipolarity, ultimately leading to digital civilization squeezing out and replacing natural civilization. As Hawking cautioned, "Humanity should be wary of the threats posed by AI development. Once AI breaks free and redesigns itself at an accelerating rate, humans, constrained by prolonged biological evolution, will be unable to compete and will thus be superseded." [15] Therefore, proposing to construct Eastern aesthetics based on the modern civilization of the Chinese nation carries macro significance not only to alert people to the ideological risks of a Western versus Eastern dichotomy inherent in Western technology but also to counteract the mechanical rationality of digital civilization with humanity's most primal aesthetic sensibility. It aims to combat the trend of civilization's unipolarity with the grand vision of a "community of shared human destiny," achieving a detachment from and conversion to the material world through the ultimate pursuit of the artistic spirit. Thus, it is necessary to construct Eastern aesthetics in the era of artificial intelligence with a transcendent philosophical stance and devout life aesthetics, which is also one of the purposes of this book's deep application of AI in the creation of ink painting.

To construct Eastern aesthetics in the era of artificial intelligence, it is first necessary to firmly uphold the cultural confidence of Chinese civilization. For over five thousand years, "the ideas inherent in China's excellent traditional culture—such as public-minded, regarding people as the foundation of the country, unremitting self-improvement, reform and innovation, governance by virtue, appointing people on their merits, unity between heaven and man, unceasing self-improvement, great virtue carrying great responsibilities, valuing trust and fostering peace, and being kind and neighborly—represent not only the cosmology, world view, social view, and moral view accumulated over long periods of production and life but also the essence of China's excellent traditional culture." [16] These cultural resources have not only supported the Chinese nation's material development to the present day but have also provided the people with boundless hope and courage in times of significant national existential crises. In this book, the painter have depicted numerous images of farmers, who, in traditional Chinese culture, are endowed with the qualities of enduring hardship, diligence, and humbleness. Whether it be farmers tilling the fields, craftsmen refining objects, or scholars tutoring pupils, all emphasize hard work. The Book of *Songs* contains the line, "Rise early and retire late, and be not ashamed of the life you lead," which praises diligence. In China's thousands of years of agrarian civilization, farmers, as the

---

14 Pang Jingjun, "Exploring the Roots of the Human Spiritual System in an Aesthetic Way," People's Forum-Academic Frontier, No. 10, 2017.

15 Xinxin Wang, "Stephen Hawking's Beijing speech: artificial intelligence could also be the end of human civilisation", https://www.thepaper.cn/newsDetail_forward_1672326, accessed 25 August 2023.

16 Yu Wenli, "Promoting Cultural Self-confidence and Self-strengthening in the New Era," New Horizons, No. 3, 2023.

foundation of society, play a crucial role in nurturing the nation and maintaining social stability. Thus, the farmers in this book represent the cultural backdrop and cultural symbols of the Chinese nation's reproduction and survival. The "unique concepts, wisdom, temperament, and charm add to the deep-seated confidence and pride of the Chinese people and the Chinese nation."[17] Therefore, the cultural confidence of Chinese civilization should be used to steadfastly reinforce the artistic confidence in oriental aesthetics.

To construct Eastern aesthetics in the era of artificial intelligence, it is also essential to maintain an open and inclusive artistic attitude. "Inclusiveness" is not only a prominent characteristic of Chinese civilization and Eastern aesthetics, but it is also an important means to enrich Eastern aesthetics continually, against the backdrop of continuously upgrading AI technology and the gradual formation of digital civilization. Before the advent of digital civilization, numerous philosophers and ideologist from both East and West profoundly discussed the importance of openness and inclusiveness to civilization, culture, and aesthetics. Liang Qichao, an ideologist from late Qing Dynasty, believed: "To take the civilization of the West to expand my own, and to use my civilization to supplement that of the West, merging them into a new civilization."[18] British philosopher Bertrand Russell also recognized the positive implications of the inclusiveness between Eastern and Western cultures for the long-term development of civilization: "They can learn indispensable practical efficiency from us; and we can learn some thoughtful wisdom from them."[19] Therefore, openness and inclusiveness is "the intrinsic driving forces behind the advancement of human civilization; no civilization can develop in isolation while completely eschewing heterogeneous civilizations."[20] In the current era of digital civilization based on artificial intelligence technology, creative methods and digital styles such as big data management, algorithmic art, singularity art, interactive art, nano-art, 3D printing art, automated generation, simulation, human-machine collaboration, and even "human-machine integration" will gradually mature and become widespread. The inclusiveness of Eastern aesthetics ensures that it will embrace different cultural features and creative methods with a proactive attitude, integrating them into its own cultural veins, thus truly invigorating the vitality of ecological civilization of Eastern aesthetic.

To construct Eastern aesthetics in the age of artificial intelligence, it is essential to continuously advance the preservation of integrity and innovation within Chinese civilization itself. Chinese civilization forms the cultural foundation of Eastern aesthetics, and its ability to uphold integrity while innovating signifies

---

17 Xi Jinping's Speech at the Opening Ceremony of the 10th Session of the China Federation of Literature and the 9th Session of the China Federation of Writers, Beijing: People's Publishing House, 2016 edition, p. 4.

18 Liang Qichao, Drinking Ice Room Collection - Monographs, Book 23, Beijing: Zhonghua Shuju, 1989 edition, p. 55.

19 [English] Bertrand Russell, The Problem of China, translated by Qin Yue, Shanghai: Xue Lin Publishing House, 1996 edition, pp. 146-156.

20 Liu Zhigang, "From "Clash of Civilisations Theory" to Community of Human Fate: The Different Logic of Chinese and Western Approaches to the Clash of Civilisations", Academia 2021, p. 207.

that Eastern aesthetics possesses an infinite cultural dynamism. The concept of "upholding integrity" entails maintaining the guiding position of Marxist literary and artistic theories in the realm of art, preserving the righteousness of creatively transforming and developing excellent traditional Chinese culture, and safeguarding the standards of Chinese aesthetic fashion and the spirit of Chinese aesthetics[21], In this catalog, traditional ink painting creation relies on the manual skills and personal experiences of artists, while the introduction of generative artificial intelligence diversifies the artistic creation process. Artists set algorithm parameters, allowing AI to learn and simulate the style of ink painting, thus creating works with unique styles. This process changes the relationship between artists and their works and the audience's understanding of the art creation process, thereby achieving a balance of preserving integrity and innovating within the creation of ink paintings.

Conclusion

The age of artificial intelligence has arrived, bringing with it not only a series of production issues in the realm of art but, more importantly, it carries an inherent cultural crisis affecting the totality of human civilization. Domestically, there has been awareness of the potential national cultural security issues that generative artificial intelligence could cause, with political advisors calling for "support for the standardized development and application of domestic ChatGPT."[22] This book also intends to contemplate from a philosophical height of human civilization how to find balance between Eastern and Western aesthetics within the creative process of ink painting, thereby addressing the future issues of ecological civilization in the age of artificial intelligence. Constructing a rich and distinctive Eastern aesthetic is an essential cultural matter. To build Eastern aesthetics in the age of artificial intelligence, one must first use Chinese civilization as the foundation and accurately identify the five cultural essences of continuity, innovation, unity, inclusiveness, and peacefulness in Eastern aesthetics. Secondly, cultural confidence, openness, and upholding integrity while innovating should be adopted as specific pathways for Eastern aesthetics, addressing the ideological contrasts between East and West and the continuation of human civilization. On the eve of widespread application of AI technology in artistic creation, we must not only actively embrace and deeply apply artificial intelligence but also dare to take our era's mission, highlighting the intentions of Eastern aesthetics with excellent works of art, promoting the spirit of Eastern aesthetics through art forms that deeply apply contemporary AI technology, and thereby fostering a better balance between human and digital civilization.

---

21 Duan Jifang, "On the "Shouzheng Innovation" of Literary and Art Criticism with Chinese Characteristics: Theory, Path, and Tasks," Social Science Series, No. 4, 2022.
22 Zhimin Gao, "Nine members of the National Committee of the Chinese People's Political Consultative Conference (CPPCC) jointly call for: support for regulating the research, development and application of domestically produced ChatGPT", http://www.rmzxb.com.cn/c/2023-05-24/3350234.shtml, accessed on 29 August 2023.

后记

重要的是培养对中国审美的"判断力"

在本书的前言部分，我向自己，向我的创作团队，也向各位读者提出了一系列围绕"人工智能与人类艺术的关系""人工智能是否可以理解美、建构美、区分美"等的问题，并对未来中国人工智能在艺术领域的发展提出了"人机协同，讲好中国故事"的美好愿景……这些理念似乎都将讨论的重心置于人工智能，那么相应的，我们更应思考人类在艺术创作领域应当坚守的又是什么呢？

我想我们应当坚守对本时代、本民族艺术文化精神的继承与发扬，而在万物互联、媒介融合的当下，重要的是培养我们自身对中国审美敏锐且精准的"判断力"。这也是我在此书中将人为绘制的画与机械生成的画"并置""拼贴"于一处的原因，我希望广大读者能够在这一直观的对比中感受到其中的差异，以避免在眼球经济盛行的当下一味跟风所谓的"潮流"，而逐渐遗失掉本土文化艺术的底色。既要使人工智能"为我所用"，又谨防成为其"吹鼓手"。对技术的发展保持审慎的乐观，进而思考在人类与算法互相补足、携手共进的未来，我们在艺术创作领域所无法被机器取代的原创力究竟从何生成，我们又当如何建构带有华夏文化属性与审美思维的数据库，如何精进算法的底层逻辑，以生成真正能够彰显中国审美之灵韵的素材，进而更好地辅助当代文艺工作者完成更为高效、更为优质、更具创造力与感染力的艺术创作，我想这是我们在视觉文化领域以人工智能"讲好中国故事"的基础。

2024

实验水墨
绘画材料：宣纸板，水墨

Experimental Ink Wash
Painting
Artwork: Rice paper board, Ink
33cm×46cm
2023

What matters is cultivating a "judgment" of Chinese aesthetics.

In the preface of this book, I raised a series of questions around "the relationship between artificial intelligence and human art" "whether artificial intelligence can understand beauty, construct beauty, and distinguish beauty" and I also proposed a vision for the future development of artificial intelligence in the field of art in China, I think we should embrace human-machine collaboration in order to better tell contemporary Chinese stories ... These ideas seem to focus on artificial intelligence, so correspondingly, what we should think more about is what humans should adhere to in the field of artistic creation when we choose to collaborate with artificial intelligence?

I believe we should adhere to the inheritance and promotion of the artistic and cultural spirit of our time and our nation. In the current era of interconnectedness and mass media convergence, it is important to cultivate a sharp and precise "judgment" of Chinese aesthetics. This is also why I juxtapose and collage hand drawn paintings with mechanically generated ones in this book. I hope readers can feel the differences in this intuitive comparison, avoiding blindly following the so-called "trends" prevalent in the era of "eyeball economy", and gradually losing our art traditions. We should make artificial intelligence "work for us" while guarding against becoming its "cheerleaders." We should maintain a rather cautious optimism about the development of technology and keep coming up with better working flows where humans and algorithms complement each other and work together. Where does the originality that cannot be replaced by machines come from in the field of artistic creation? How should we construct databases with Chinese cultural attributes and aesthetic patterns? How can we refine the underlying logic of algorithms to generate materials that truly embody the spirit of Chinese aesthetics? This will better assist contemporary artistic creators in completing more efficient, higher-quality, more creative, and more infectious artistic creations. I believe this is the foundation for us to tell Chinese stories better in the future in the field of visual culture using artificial intelligence.

AI 衍生数字绘画
由亦心闪绘生成

Digital painting based on the painting
Generated by AI flash painting of the company YiXin
928px × 1232px
2023

AI 衍生数字绘画
由亦心闪绘生成

Digital painting based on the painting
Generated by AI flash painting of the company YiXin
928px × 1232px
2023

AI 绘画：当代水墨艺术的"正发生" | 351

图书在版编目(CIP)数据

AI绘画：当代水墨艺术的"正发生" / 凡悲鲁著.
— 北京：海洋出版社, 2024.6
ISBN 978-7-5210-1267-5

Ⅰ.①A… Ⅱ.①凡… Ⅲ.①图像处理软件 Ⅳ.①TP391.413

中国国家版本馆CIP数据核字(2024)第111130号

北京电影学院数字影视动画创作教育部工程研究中心
动画新媒体技术北京市重点实验室
数字电影技术与艺术重点实验室—动画分实验室
资助出版

装帧设计：祖若曦
版式设计：祖若曦　金子簃　杨小汀　任芷蕙
采　　编：霍笑妍
AI 生 成：刘昌伟　韦祖兴　安阳阳　傅　岩　杨　亮
　　　　　张文正　张　元　赵伟然　韩泽宇
宣　　传：影　子
统　　筹：Sue
鸣谢北京亦心科技有限公司提供技术支持

责任编辑：赵　武
排　　版：申　彪
责任印制：安　淼

海洋出版社　出版发行
http://www.oceanpress.com.cn
北京市海淀区大慧寺路 8 号　邮编：100081
侨友印刷（河北）有限公司印刷
2024年6月第1版　　2024年6月第1次印刷
开本：889mm×1194mm　1 / 12　印张：30
字数：300千字　定价：680.00 元

发行部：010-62100090　总编室：010-62100034
海洋版图书印、装错误可随时退换